COMPLETE GUIDE TO TELEPHONE EQUIPMENT TROUBLESHOOTING AND REPAIR

OTHER BOOKS BY JOHN D. LENK

- (159955) COMPLETE GUIDE TO COMPACT DISC (CD) PLAYER TROUBLESHOOTING AND REPAIR (1986)
- (160359) COMPLETE GUIDE TO MODERN VCR TROUBLESHOOTING AND REPAIR (1985)
- (160813) COMPLETE GUIDE TO LASER/VIDEODISC PLAYER TROUBLESHOOTING AND REPAIR (1985)
- (160820) COMPLETE GUIDE TO VIDEOCASSETTE RECORDER OPERATION AND SERVICING (1983)
- (392473) A HOBBYIST'S GUIDE TO COMPUTER EXPERIMENTATION (1985)
- (372391) HANDBOOK OF ADVANCED TROUBLESHOOTING (1983)
- (377317) HANDBOOK OF DATA COMMUNICATIONS (1984)
- (380519) HANDBOOK OF MICROCOMPUTER BASED INSTRUMENTATION AND CONTROLS (1984)
- (381666) HANDBOOK OF SIMPLIFIED COMMERCIAL AND INDUSTRIAL WIRING DESIGN (1984)

To order, write: Steven T. Landis
Book Distribution Center
Route 59 at Brook Hill Drive
West Nyack, New York 10995
or call: (201)767-5049 through 5053

COMPLETE GUIDE TO TELEPHONE EQUIPMENT TROUBLESHOOTING AND REPAIR

John D. Lenk

Consulting Technical Writer

Prentice-Hall, Inc., Englewood Cliffs, N.J. 07632

Library of Congress Cataloging-in-Publication Data

LENK, JOHN D. (date)
   Complete guide to telephone equipment troubleshooting and repair.

   Includes index.
   1. Telephone—Equipment and supplies—Maintenance and repair. I. Title.
TK6471.L46 1987   621.385   86-5043
ISBN 0-13-160797-9

Editorial/production supervision and
   interior design: Erica Orloff and Natalie Brenner
Manufacturing buyer: Gordon Osbourne

© 1987 by Prentice-Hall, Inc.
A Division of Simon & Schuster
Englewood Cliffs, New Jersey 07632

All rights reserved. No part of this book may be
reproduced, in any form or by any means,
without permission in writing from the publisher.

Printed in the United States of America
10 9 8 7 6 5 4 3 2 1

ISBN 0-13-160797-9 025

Prentice-Hall International (UK) Limited, *London*
Prentice-Hall of Australia Pty. Limited, *Sydney*
Prentice-Hall Canada Inc., *Toronto*
Prentice-Hall Hispanoamericana, S.A., *Mexico*
Prentice-Hall of India Private Limited, *New Delhi*
Prentice-Hall of Japan, Inc., *Tokyo*
Prentice-Hall of Southeast Asia Pte. Ltd., *Singapore*
Editora Prentice-Hall do Brasil, Ltda., *Rio de Janeiro*

To my Irene: I am so glad that you and I became US!
To Lambie: A very special FUZZY PEOPLE!

CONTENTS

PREFACE xi

1 INTRODUCTION TO CONSUMER TELEPHONE EQUIPMENT .. 1

    1-1. Standard telephone characteristics and functions, *2*
    1-2. Basic electronic telephone circuits, *5*
    1-3. Basic cordless telephone circuits, *10*
    1-4. Problems in servicing telephone equipment, *15*

2 THE BASICS OF TELEPHONE EQUIPMENT TROUBLESHOOTING AND REPAIR 17

    2-1. Safety precautions in telephone equipment service, *17*
    2-2. Test equipment for telephone service, *22*
    2-3. B&K-Precision Model 1050 Telephone Analyzer, *23*
    2-4. B&K-Precision Model 1047 Cordless Telephone Tester, *56*
    2-5. B&K-Precision Model 1045 Telephone Product Tester, *57*
    2-6. B&K-Precision Model 1042 Telephone Line Analyzer, *67*
    2-7. The basic troubleshooting approach, *75*

3 THE BASICS OF TROUBLESHOOTING CORDED TELEPHONES 79

3-1. Testing the telephone line, 79
3-2. Testing cords, 80
3-3. Testing the ringer circuits, 81
3-4. Testing the dialing circuits, 85
3-5. Testing the audio circuits, 90

4 THE BASICS OF TROUBLESHOOTING CORDLESS TELEPHONES 95

4-1. Testing the telephone line and base-unit cords, 95
4-2. Preliminary checkout, 96
4-3. Ring-circuit basic troubleshooting, 97
4-4. Dial-tone circuit basic troubleshooting, 102
4-5. Dial-circuit basic troubleshooting, 107
4-6. Voice-circuit basic troubleshooting, 111
4-7. Short-range problems, 114
4-8. Problems in digitally coded cordless telephones, 116

5 CORDLESS TELEPHONE TROUBLESHOOTING 119

5-1. Introduction to cordless telephone circuits, 120
5-2. Base-unit incoming communications circuits, 122
5-3. Base-unit outgoing communications circuits, 125
5-4. Portable-unit communications circuits, 133
5-5. Adjustment procedures, 145
5-6. Introduction to cordless telephone troubleshooting, 154
5-7. Base unit dead, power indicator does not turn on, 155
5-8. Telephone line cannot be seized, 155
5-9. Telephone does not ring, 159
5-10. Voice from portable unit cannot be heard on telephone line, 161
5-11. Portable unit rings but calling party cannot be heard, 163
5-12. Base unit does not pass through dial signals, 164
5-13. No paging buzzer (call signal) when CALL button is pressed, 164

Contents ix

    5-14. IN-USE indicator does not turn on, *164*
    5-15. CHARGE indicator does not turn on, *165*
    5-16. Portable unit dead, *165*
    5-17. Battery does not charge with portable unit in cradle, *166*
    5-18. Battery does not charge from external source, *166*
    5-19. BATT LOW indicator does not turn on and off properly, *167*
    5-20. TALK indicator does not turn on, *168*
    5-21. Portable unit does not ring, *168*
    5-22. Telephone line cannot be seized or base unit activated, *169*
    5-23. No dialing or redial functions, *172*
    5-24. Telephone rings but no sound on portable-unit speaker, *175*
    5-25. Telephone rings but voice from portable unit cannot be heard, *176*
    5-26. Flash (call-waiting) not operative, *178*
    5-27. Automatic standby function not operative, *178*
    5-28. No keystroke pulse generated when dial keys are pressed, *179*

6 MODEM AND TELEPHONE INTERFACE TROUBLESHOOTING 181

    6-1. Basic modem system, *181*
    6-2. Modem I troubleshooting, *189*
    6-3. Modem II troubleshooting, *201*
    6-4. Telephone Interface II (acoustic coupler) troubleshooting, *221*

INDEX **235**

PREFACE

The main purpose of this book is to provide a simplified system of troubleshooting and repair for the many types of telephone equipment. It is virtually impossible in one book to cover detailed troubleshooting and repair for all telephone equipment. Similarly, it is impractical to attempt such comprehensive coverage, since rapid technological advances soon make such a book's detail obsolete.

To overcome this problem, this book concentrates on a basic approach to telephone equipment service, an approach that can be applied to any telephone device (equipment now in use and equipment to be manufactured in the future). The approach here is based on the techniques found in the author's best-selling *Handbook of Practical Solid-State Troubleshooting, Handbook of Advanced Troubleshooting, Complete Guide to Videocassette Recorder Operation and Servicing, Complete Guide to Laser/Videodisc Player Troubleshooting and Repair, Complete Guide to Modern VCR Troubleshooting and Repair,* and *Complete Guide to Compact Disc (CD) Player Troubleshooting and Repair.*

Chapter 1 is devoted to the basics of telephone equipment, including the relationship to the telephone lines and exchanges. With the basics established, the chapter then describes the technical characteristics for the two most popular forms of telephone equipment: the electronic corded telephone and the cordless telephone. The chapter concludes with a summary of problems in telephone equipment service.

Chapter 2 discusses the basic approaches for troubleshooting and repair of telephone equipment. The chapter covers such areas as safety precautions, test equipment, tools, and basic troubleshooting, and notes that apply to telephone equipment of all types. The troubleshooting procedures for specific types of telephone equipment are covered in the related chapter.

Chapter 3 describes the basics of troubleshooting corded-telephone equipment. As discussed throughout the book, the first step in troubleshooting is to test the instrument against known standards. Chapter 2 describes a complete set of tests for typical corded-telephone equipment using specialized test equipment. Chapter 3 describes how to perform similar tests without specialized test equipment. By comparing the procedures of the two chapters, the reader will quickly realize the advantages of specialized test equipment, particularly if they plan on servicing telephone equipment regularly.

Chapter 4 provides coverage similar to that given in Chapter 3, but for cordless telephones of all types.

Chapter 5 describes troubleshooting and service notes for a cross section of cordless telephone equipment. The chapter is divided into three parts: circuit descriptions, test/adjustment procedures, and circuit-by-circuit troubleshooting. By studying the circuits found in this chapter, the reader should have no difficulty in understanding the schematic and block diagrams of similar telephones. Because adjustments are closely related to troubleshooting, the chapter describes typical adjustment procedures for cordless telephones. Using the examples, the reader should be able to relate the procedures to a similar set of adjustments on most cordless telephones. Where it is not obvious, the purpose of the adjustment procedure is also described. The waveforms and signals measured at various points during adjustment are also included. By studying these waveforms, the reader can identify typical signals found in most cordless telephones, even though the signals may appear at different points.

With adjustments well established, the chapter then describes *circuit-by-circuit troubleshooting* for a cross section of cordless telephones. This ciruit-by-circuit approach is based on *failure or trouble symptoms* and represents the combined experience and knowledge of many telephone equipment service specialists and managers.

Chapter 6 provides coverage similar to that given in Chapter 5 (circuit descriptions, adjustments, and circuit-by-circuit troubleshooting), but for modems, telephone interface equipment, and acoustic couplers.

Many professionals have contributed their talents and knowledge to the preparation of this book. The author gratefully acknowledges that the tremendous effort to make this book such a comprehensive work is impossible for one person, and wishes to thank all who have contributed directly and indirectly.

The author wishes to give special thanks to the following: Martin Pludé, Christopher Kite, and Bob Carlson of B&K-Precision Dynascan Corporation; Dave Gunzel, Hy Seigel, and Ted Rosenberg of Radio Shack.

The author extends his gratitude to Matt Fox, Diana Spina, Tim McEwen, Rosemary Mahoney, Andy O'Hearn, Melissa Halverstadt, Leon Liguori, Greg Burnell, Dave Boelio, Tony Caruso, Hank Kennedy, John Davis, Barbara Cassel, Jerry Slawney, Art Rittenberg, Ellen Denning, Beverly Vill, Mary O'Brien, Irene

Springer, Nancy Bauer, Karen Fortgang, Lisa Shulz, Kathy Pavelec, Nina Seigelstein, and Armond Fangschlyster of Prentice-Hall, and Marie Barlettano of PH International. Their faith in the author has given him encouragement, and their editorial/marketing expertise has made many of the author's books best-sellers. The credit must go to them. The author also wishes to thank Joseph A. Labok of Los Angeles Valley College for his help and encouragement throughout the years.

JOHN D. LENK

COMPLETE GUIDE TO TELEPHONE EQUIPMENT TROUBLESHOOTING AND REPAIR

1

INTRODUCTION TO CONSUMER TELEPHONE EQUIPMENT

This chapter is devoted to the basics of telephone equipment. Although there are many types of consumer telephone equipment (standard, electronic feature, answering machines, special feature, etc.), there are only two basic groups: corded and cordless. We cover the basics of both groups (briefly) in this chapter. Full details of circuit operation for a particular type of telephone equipment are covered in the related chapters. However, it is helpful if you read the basic functions described here, particularly if you are new to telephone equipment service. We all use the telephone, but few of us know the basic functions.

The techniques required to service telephone equipment are essentially the same as for other electronic equipment. You must know how to use test equipment, make observations, analyze test results, use common hand tools, and so on. If you are not familiar with these, and you plan to service telephone equipment, you can be in considerable trouble rather quickly. Keep the following in mind. If you make a mistake and destroy a TV set or audio system during service, one customer will be most unhappy. If you make a mistake that blows circuits at the telephone exchange, many, many people will be unhappy!

To prevent such an unspeakable possibility, the author *strongly recommends* that you use the highly specialized test equipment readily available for consumer telephone troubleshooting and repair. Chapter 2 discusses off-the-shelf instruments for testing and adjusting virtually all types of consumer telephone equipment. Much of this book is devoted to the operation and use of such commercial test equipment. In addition to making life much easier for the service technician, proper use of the special test equipment eliminates the probability of incurring the wrath of your local telephone company.

Do not let any of this scare you off. You should have no difficulty in understanding the test equipment or in troubleshooting the telephone equipment circuits described here. One possible exception is if you do not understand electronic troubleshooting. In that event, the problem can be cured by reading the author's best-selling troubleshooting books: *Handbook of Practical Solid-State Troubleshooting, Handbook of Basic Electronic Troubleshooting,* and *Handbook of Advanced Troubleshooting.*

Now let us start with the basics of consumer telephone operation.

1-1. STANDARD TELEPHONE CHARACTERISTICS AND FUNCTIONS

So that all telephone equipment is compatible (and usually interchangeable) with other telephones and telephone exchanges, it is essential that all telephones have certain standard characteristics. Figure 1-1 shows the basic characteristics involved for both the telephone and exchange. Before we get into these characteristics, let us consider the basic functions of any telephone system.

Obviously, a telephone system must be capable of *transmitting* and *receiving* voice or other sounds. The system must also be capable of *addressing* or indicating the person or location you wish to call. Generally, addressing is referred to as the *dialing* function. Next, the system must be able to *alert* the person or station addressed. Generally, the alerting function is done by *ringing* the

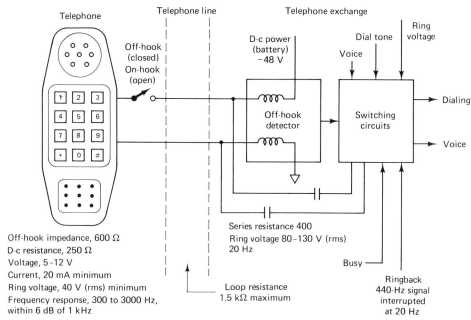

FIGURE 1-1. Basic characteristics for both telephones and telephone exchanges

telephone. Finally, the system must *supervise* all of these functions and continuously scan the circuits to determine when someone wishes to place a call. Lifting the telephone *off-hook* indicates this request-to-call condition to the switching equipment at the telephone exchange. The exchange responds to the request by transmitting a *dial tone* back to the off-hook telephone.

1-1.1 The Two-Wire Circuit

Although you may find four wires in some telephone cords and outlets, the standard telephone system uses only two wires (or one pair), as shown in Fig. 1-1. The extra wires (second pair) in a telephone cord can be put to various uses, such as providing power to a dial light, but only two wires are required to perform all of the basic functions just discussed. (This may lead to some confusion, particularly in those forms of communication where four-wire circuits are used. In such four-wire circuits, transmit-voice is carried by one pair of wires, and receive-voice is carried by the other pair.) In a standard two-wire telephone circuit, voice audio in both directions, d-c power for operating the telephone circuits, dial tone, dialing signals, and ringing are all carried on a single pair of wires.

1-1.2 Off-Hook Impedance and Bandwidth or Frequency Response

The typical off-hook impedance of a standard telephone is about 600 Ω at frequencies between 300 and 3000 Hz. Most standard telephones have a frequency response of about 300 to 3000 Hz, within 6 dB of the 1-kHz response. The typical d-c resistance of an off-hook telephone is about 250 Ω.

1-1.3 Battery Voltage (D-C Operating Voltage)

As shown in Fig. 1-1, the telephone exchange continuously applies a d-c voltage to the telephone line through a series impedance. Typically, this voltage is -48 V and is produced by a giant bank of batteries connected in parallel. Batteries are used to provide telephone operation in case of power failure.

When the telephone is on-hook, the telephone appears as an open circuit and no current flows on the line. When the telephone is off-hook, line current flows and powers the telephone.

A standard telephone exchange is required to supply at least 20 mA of line current. (20 mA is the value originally required for satisfactory operation of carbon microphones, no longer used in modern telephones. Those familiar with computers and modems will also recognize 20 mA as the standard current for telephone and teletype line connections.)

The *loop resistance* of a long telephone line may be up to 1500 Ω. Loop resistance refers to the resistance from the exchange to the telephone, and from

4   Introduction to Consumer Telephone Equipment

the telephone back to the exchange. When the full 1500 Ω value is present, the line current must still be 20 mA. For shorter telephone lines, the available line current may be as high as 120 mA. However, most telephones have circuits to automatically shunt excess line current. Although about 48 V is available at the exchange, the voltage across an off-hook telephone is usually 5 to 12 V.

The standard polarity for the d-c operating voltage (battery voltage) is negative to the "ring" and return to the "tip." ("Ring" and "tip" are telephone terms applied to the concentric ring and tip portions of a telephone jack and have nothing to do with ringing of the telephone.)

1-1.4   Voice Signals

When voice is present, *current varitions are superimposed* on the d-c operating voltage. Typically, a voltage regulator within the telephone provides a constant voltage to the telephone circuits even though the line current is varied by voice signals.

1-1.5   The Ringing Voltage

At this point you may be wondering how the exchange rings a particular telephone when the telephone is in the on-hook or open condition and no d-c operating voltage is applied to the telephone circuits. (If you are not wondering this, you may not be paying attention!) The exchange applies a 20-Hz sine wave *ringing voltage* of about 80 to 130 V (rms). Typically, the ringing voltage is 100 V (or 280 V peak to peak). Telephones must be capable of ringing properly with a ringing voltage of 40 V (which is the value the ringing voltage may drop to over a long telephone line with 1500 Ω impedance).

The ringing voltage or signal is applied to the telephone line only when the exchange detects a high impedance (when the telephone is on-hook). The ringing is usually 2 to 2.5 seconds on, and 3.5 to 4 seconds off, for a 6-second repetition cycle.

Although the standard frequency for the ringing signal is 20 Hz, additional frequencies may be used in some party-line applications. In those cases, telephone ringers with corresponding resonant frequencies assure that only one telephone rings.

1-1.6   The Dialing Systems

Dialing is done by means of *pulses* (for rotary or non-tone-dial telephones) or *tone pairs* (for tone-dial telephones).

On *rotary-dial telephones*, the pulses are produced by opening and closing a set of contacts. For *pushbutton pulse-dial telephones*, the pulses are produced by turning a switching circuit on and off.

With Touch Tone ("Touch Tone" is a registered trademark of AT&T) and other compatible telephones, the dialing tones are produced by oscillators that generate tones. The tones are mixed in pairs to form each digit. The 3 × 4 matrix layout of the dialing keypad (shown in Fig. 1-1) selects one "row" oscillator and one "column" oscillator for each digit. These are commonly known as DTMF (Dual-Tone Multi Frequency) frequencies. The standard DTMF dialing-tone frequencies are shown in Fig. 1-2.

1-2. BASIC ELECTRONIC TELEPHONE CIRCUITS

Figure 1-3 is the block diagram of a typical electronic telephone. Let us go through the overall functions of the circuits shown before we describe the details.

The ringing signal from the exchange is applied to an *electronic ringer circuit* through capacitive coupling. The electronic ringer consists of an oscillator and buzzer which operate at a rate of about 2.8 kHz, interrupted at a rate of 20 Hz. This interruption causes the "chirping" sound that is heard when a telephone rings. The signals are applied only to the buzzer while the ringing signal is present. So, if the ringing signal is on for 2 seconds and off for 4 seconds, the chirping is heard for 2 seconds (at 4-second intervals).

For standard desk telephones (those with a bell rather than an electronic

Digit or Symbol	Frequency Pairs (Hz)
1	697 1209
2	697 1336
3	697 1477
4	770 1209
5	770 1336
6	770 1477
7	852 1209
8	852 1336
9	852 1477
0	941 1336
*	941 1209
#	941 1477

FIGURE 1-2. Standard DTMF dialing tone frequencies

6   Introduction to Consumer Telephone Equipment

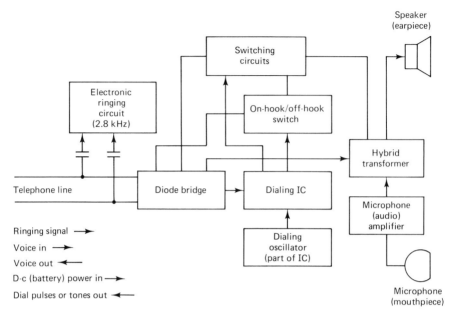

FIGURE 1-3. Block diagram of a typical electronic telephone

ringer), the ringing signal is applied to an electromechanical bell through capacitive coupling. The bell then rings at its own mechanical resonant frequency of 20 Hz.

Capacitive coupling is used with both electronic and nonelectronic telephones. In either case, the ringer must not offer a d-c path for line current when the telephone is on-hook.

D-c (battery) power, incoming and outgoing audio signals, and dialing signals are applied through a *diode bridge*, an *on-hook/off-hook switch*, and a *switching circuit*. The diode bridge serves as an automatic polarity corrector for the electronic circuits in the telephone, providing the proper polarity even if the polarity of the telephone line connection becomes reversed (from the normal -48 V). When the telephone is off-hook, the switching circuit applies the d-c power to the audio or microphone amplifier.

Note that the *dialing IC* gets d-c power directly from the diode bridge, bypassing the on-hook/off-hook switch. This is necessary with *redial* and *memory telephones* because the dialing IC must always have power to store the redial information in the memory.

When the telephone is dialed, an oscillator in the dialing IC generates a certain number of pulses for each digit. These pulses cause the switching circuit to emit the proper pulse train for each digit. The pulse train is fed through the on-hook/off-hook switch and the diode bridge onto the telephone line. The dialing pulses are fed through the telephone line to the exchange, where the pulses are decoded to make the proper connection.

When you talk, the outgoing audio is fed into an *audio amplifier* from the microphone. This amplified signal is fed to the primary of the *hybrid transformer*, which superimposes the signal on the d-c operating voltage. The voice signal is passed through the switching circuit, on-hook/off-hook switch, and diode bridge out to the telephone line. The signal is then connected to the telephone at the other end of the line by the exchange.

The voice signal is also fed to the secondary of the hybrid transformer, which feeds the signal at a low level to the speaker so that you hear your own voice in the earpiece as you speak. This is known as *sidetone*. Without this sidetone, the conversation would sound unnatural. Also note that if the sidetone level is too high, the person speaking will speak too softly, and vice versa.

Incoming audio is fed by the exchange though the line and into the telephone. This signal passes through the diode bridge, on-hook/off-hook switch, switching circuit, and hybrid transformer to the speaker.

1-2.1 Typical Ringer Circuit for an Electronic Telephone

Figure 1-4 shows the ringer circuit of an electronic telephone. The ring buzzer BZ1 and oscillator Q101 are coupled to the telephone line through C1. Capacitive coupling is used to prevent the d-c voltage from turning Q101 on. A 27-V zener diode ZD101 is also connected between the oscillator and telephone line to prevent small signals, such as dial tones or audio, from triggering Q101. Note that Q101 can be disabled to prevent ringing when power to Q101 is interrupted by opening ringer on/off switch SW102.

The ringing signal (of at least 40 Vrms at 20 Hz) exceeds the ZD101 zener

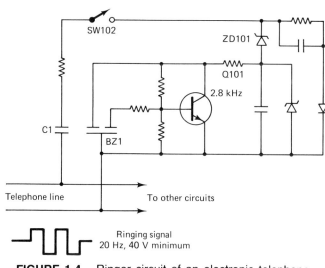

FIGURE 1-4. Ringer circuit of an electronic telephone

8 Introduction to Consumer Telephone Equipment

voltage turn-on point, and provides power to Q101, which oscillates at approximately 2.8 kHz. Power from the ring signal is essentially a 20-Hz square wave that interrrupts the 2.8-kHz oscillations at a 20-Hz rate.

1-2.2 Typical Dialing Circuit for an Electronic Telephone

Figure 1-5 shows the dialing circuit of an electronic telephone. As shown, power for the dialing circuit is taken from the telephone line (-48 V) and applied through diode bridge D101–D104 to the keypad, dialer IC101, and switching

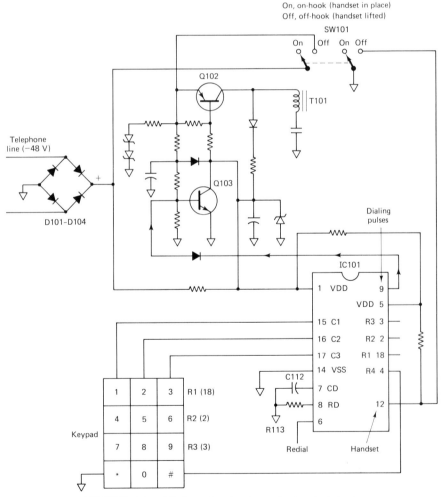

FIGURE 1-5. Dialing circuit of an electronic telephone

circuit Q102/Q103. The diode bridge keeps the polarity of the power applied to the dialing circuit constant, even if the polarity of the -48 V provided by the exchange is reversed.

The dialing keypad is connected to IC101, where each digit that has been dialed is read. Since Fig. 1-5 shows the circuit of a pulse-dial telephone, rather than a tone-dial instrument, IC101 releases a corresponding number of pulses for each digit. A storage circuit within IC101 allows the digits to be dialed faster than they are clocked out. C112 and R113 are a timing circuit used to control IC101 so that the dialing pulses are clocked at a steady rate.

The dialing pulses from IC101 are applied to Q102/Q103, which turn on and off. The pulses at the emitter of Q102 are applied to the telephone line and exchange through the hook on/off switch SW101 and diode bridge D101–D104. The collector of Q102 is also connected to hybrid transformer T101 as discussed in Sec. 1-2.3. Note that IC101 produces the dialing pulses (at pin 9) only when pin 12 is returned to ground through the other contacts of hook on/off switch SW101, in the off-hook (handset lifted) condition. Also, when the telephone is hung up (on-hook, handset replaced), the last number dialed is latched into memory within IC101 so that the number can be released when a redial button is pressed. Not all electronic telephones have a redial function.

1-2.3 Typical Audio Circuit for an Electronic Telephone

Figure 1-6 shows the audio circuit for an electronic telephone. Note that power for the audio circuit is taken from the telephone line (-48 V) through diode bridge D101–D104, SW101, Q102, and the primary of hybrid transformer T101.

When you speak into the microphone (on the handset), the signal is amplified by Q104/Q105. The primary of T101 acts as the audio load for amplifier circuit Q104/Q105, which varies current through the primary of T101 at the audio rate. These current variations are applied to the telephone line through Q102, hook on/off switch SW101, and the diode bridge D101–D104. The variations appear at the voice transformer of the telephone exchange and are applied to the telephone at the receiving end as incoming audio. ZD105 keeps the operating voltages for Q104/Q105 constant, in spite of the current variations.

Incoming audio is in the form of current variations on the telephone line. These current variations are fed through the diode bridge, SW101, Q102, and T101 to the audio circuit. The audio signals are coupled through the secondary of T101 to the speaker (in the earpiece of the handset). Diodes D108 and D109 protect the speaker from voltage spikes that might be present by clipping off any signal above the forward voltage drop of the diodes. Note that the volume of the audio to the speaker can be controlled by volume hi/lo switch SW103.

Note, too, that some of the amplified microphone audio from Q104/Q105 is also coupled into the secondary of T101 and applied to the speaker, along with the incoming audio. This microphone audio is the sidetone described in Sec. 1-2.

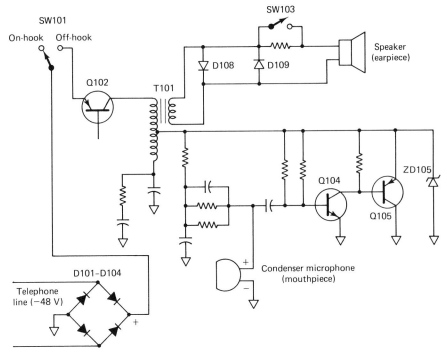

FIGURE 1-6. Audio circuit for an electronic telephone

1-3. BASIC CORDLESS TELEPHONE CIRCUITS

Cordless telephones have all the circuits and functions of the basic electronic telephone described in Sec. 1-2, plus a *two-way radio communications link*. This link covers not only voice communications, but the dialing, ringing, and hook on/off switch control functions.

Figure 1-7 shows typical cordless telephone operation. As shown, cordless telephones are composed of two separate instruments: a *base unit* and a *portable unit*. The two units are linked together using full-duplex FM radio. (In this case, full-duplex means that the base-to-portable link and portable-to-base link can operate simultaneously.) Both the base and portable units have a transmitter and a receiver. The base unit is connected directly to the telephone line and serves as the link between the portable unit and the telephone line.

Figures 1-8 and 1-9 are block diagrams of a typical cordless telephone portable unit and a base unit, respectively. Note that the diagrams show both 1.7- and 46-MHz operation. This is because cordless telephone base-to-portable links use a carrier signal in either the 1.7- or 46-MHz frequency band. Earlier cordless telephone base-to-portable links use the 1.7-MHz band, with the *a-c power line* used as an antenna. The power line is used because a very long antenna is needed for optimum transmission of a signal at 1.7 MHz. Newer cordless telephone base-

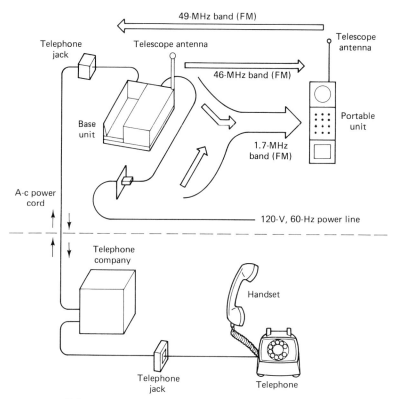

FIGURE 1-7. Typical cordless telephone operation

to-portable links use the 46-MHz band, with a *telescoping antenna*. In either case, the base unit contains a power supply (operating from the a-c line voltage) which is used to power the transmitter and receiver of the base unit and to charge the batteries of the portable unit.

Cordless telephone portable-to-base links use a carrier signal in the 49-MHz band. This signal is transmitted using the telescoping antenna (found on both portable and base units). In addition to the transmitter and receiver, portable units contain a speaker, microphone, dialing keypad, and rechargeable batteries.

1-3.1 Privacy in Cordless Telephones

Because of the limited number of cordless telephone RF channels, the same channels are used by many telephones simultaneously. This can create some obvious privacy problems. However, with the relatively short range of cordless telephones, and the use of certain techniques, a minimum of interference, false ringing, and security problems (unauthorized capturing of a telephone line) are present in cordless telephones. Let us review the basic techniques for ensuring privacy in cordless telephones.

Specific ring frequencies. False ringing is reduced by using several specific ring frequencies for the base-to-portable link. In this way, two cordless telephones with overlapping range operating on the same RF channel but with different ring frequencies do not cause ringing of the neighboring unit.

Specific guardtones or pilot signals. Use of several specific guardtones or pilot signals for the portable-to-base link reduces unauthorized use of a telephone line. The base unit does not respond to a portable unit unless the proper guardtone (called the pilot signal in some equipment) is sent by the portable. This prevents "capture" of a base unit by a nearby portable unit on the same RF channel, but with a different guardtone.

Digital coding. Early-model cordless telephones use only guardtones and ring frequencies to help prevent unauthorized capture of telephone lines and false ringing. Newer-model cordless telephones use digital coding which precedes ringing and base-unit capture. With digital coding, it is not necessary to use individual guardtones and ring frequencies. Usually, a single guardtone and ring frequency are used for all cordless telephones with digital coding. Of course, the digital coding methods vary from one cordless telephone manufacturer to another. This is why you must consult the service manual when servicing any telephone instrument.

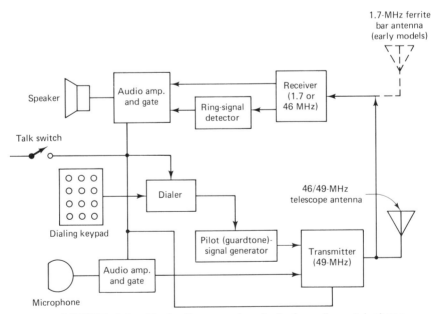

FIGURE 1-8. Block diagram of a typical cordless telephone portable unit

1-3.2 Incoming Calls on a Cordless Telephone

Between calls, the cordless telephone is in a "standby" mode, where both the base and portable transmitters are off and both receivers are awaiting incoming RF carrier signals.

When a 20-Hz ringing signal is received by the base unit (Fig. 1-9) from the telephone line, a 20-Hz ring detector turns on the base unit transmitter and a *ring-signal generator*. This generator feeds a ring signal (at one particular freuqency) to the transmitter. Typically, the ring signal is in the range 700 to 1500 Hz. Different ring frequencies help prevent false ringing, as discussed. At the transmitter, the ring signal is used to modulate a carrier signal (in either the 1.7- or 46-MHz band). The modulated carrier is then transmitted to the portable unit.

The receiver section of the portable unit (Fig. 1-8) has power if the portable unit is in either the talk or standby mode, as long as the power is turned on and the batteries are charged. The incoming signal is demodulated by the receiver and fed to the *ring-signal detector*. This detector is a filter that only passes the ring signals of one specific frequency. If the ring frequency transmitted by the base unit is correct (same as the ring-signal filter frequency), the ring signal is passed to an audio amplifier where the signal is amplified and fed to the speaker.

When the call is answered, the portable unit is switched from the standby

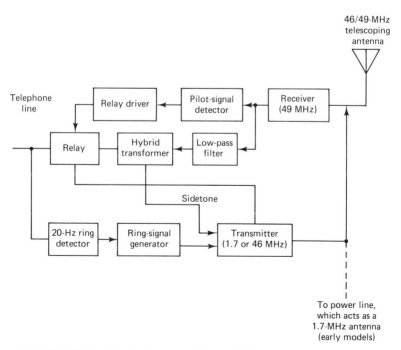

FIGURE 1-9. Block diagram of a typical cordless telephone base unit

mode to the talk mode. This disconnects the ring signal from the amplifier and turns on the RF transmitter and *pilot-signal generator*, as well as the audio amplifiers and gates. The pilot signal or guardtone, and audio from the microphone, are fed to the transmitter, where they modulate a 49-MHz carrier. This modulated signal is then transmitted over the telescoping antenna to the base unit.

The base unit (Fig. 1-9) receives and demodulates the signal, and then feeds the signal to the *pilot-signal detector* (low-pass filter). When the pilot signal or guardtone is at the correct frequency, the pilot-signal detector energizes a relay into the off-hook condition through operation of a relay driver. The relay turns on the base-unit transmitter.

The low-pass filter blocks the guardtone and feeds voice audio (300 to 3000 Hz) to the hybrid transformer. In turn, the hybrid transformer feeds the audio signal through the relay to the telephone line (when the relay is placed in the off-hook condition by a guardtone of the correct frequency). The hybrid transformer also feeds a low-level audio signal to the transmitter for sidetone (Sec. 1-2) in the portable unit.

Incoming audio from the telephone line is fed through the relay to the hybrid transformer when the relay is in the off-hook condition. This audio signal is fed to the transmitter, where it is used to modulate a carrier signal in either the 1.7- or 46-MHz band. This modulated signal is transmitted over the antenna to the portable unit.

The portable unit receiver demodulates and feeds the signal to the audio amplifier and gate. When the portable-unit switch is in the talk mode, the demodulated audio signal is amplified and fed to the speaker (earpiece).

1-3.3 Outgoing Calls on a Cordless Telephone

When a telephone call is initiated from a cordless telephone, the portable unit (Fig. 1-8) is switched from the standby mode to the talk mode. This turns on the transmitter and pilot signal or guardtone. The pilot signal is typically in the range 4 to 7 kHz, and modulates the 49-MHz carrier. The modulated signal is transmitted to the base unit over the telescoping antenna.

The signal from the portable unit is demodulated by the base-unit (Fig. 1-9) receiver and is fed to the pilot-signal detector and low-pass filter. The pilot-signal detector energizes the relay to an off-hook state (through the relay driver) if the pilot signal is at the correct frequency. The cordless telephone is now in the off-hook condition, and a telephone number can be dialed. The low-pass filter allows only signals at or below 3 kHz to pass, so the guardtone or pilot signal is rejected while the dialing and voice signals are allowed to pass.

Pulse-dial cordless. When a digit is pressed at the portable unit of a pulse-dial cordless telephone, the pilot signal is interrupted a certain number of times for each digit. The relay interrupts the off-hook condition each time the pilot signal is interrupted. This on-hook/off-hook change at the relay causes dialing pulses to go out over the telephone line. These pulses are decoded by the telephone exchange just as with a standard pulse-dial telephone, and the telephone call is routed to the telephone number dialed.

Tone-dial cordless. When a digit is pressed at the portable unit of a tone-dial cordless telephone, the dialing tones and guardtone are used to modulate the carrier. Dialing tones are in the 300- to 3000-Hz band (Fig. 1-2), while guardtones are above 3000 Hz. The modulated signal is received by the base unit, demodulated, and then (if the guardtone frequency is correct) the tones are sent out over the telephone line. At the exchange, the tones are decoded and used to route the telephone call as with a standard tone-dial telephone.

As discussed previously, in the off-hook state, both transmitters are in a full-duplex condition (to permit talking and listening simultaneously). Incoming and outgoing audio are handled as described.

When the call is completed, the cordless telephone portable unit is set back to the standby mode, and the pilot signal is discontinued. This causes the base unit relay to go back to an on-hook condition, and incoming calls can now be accepted.

1-4. PROBLEMS IN SERVICING TELEPHONE EQUIPMENT

Now that we know the basics of how telephones operate, we will get into the details of test and troubleshooting, starting with test equipment in Chapter 2. First let us consider some general problems in servicing telephone equipment. Besides the obvious problem of "tampering" with the telephone company lines, there are three areas to consider.

First, you may have difficulty in getting service literature and replacement parts for telephone equipment, although some manufacturers produce excellent service literature and maintain good stocks of replacement parts. Also, many parts are interchangeable from one manufacturer to another, since all telephones must be compatible with the telephone company. On the other hand, certain manufacturers want their telephone equipment to be returned to them, or to their authorized service shops, for all service. They do this by providing little or no service information and/or replacement parts. Of course, this problem is not unique to telephone equipment. Anyone now in the TV/video/audio service field knows this well.

Second, telephone equipment is changing rapidly. Some manufacturers see no point in providing literature and parts for outside service on equipment that will shortly be replaced by a new model. They may be right, but it leaves you (the service technician) to take the flack from the public (which is also not an uncommon problem).

Finally, you are up against the problem of "throwaway" telephone equipment, where the price of the instrument is below that of an hour or two of service (to say nothing of the cost of replacement parts). Either the customer will throw the instrument away, or you will not make money, or both. On the other hand, there are millions of telephones in use, and they will probably require service at some point in their lifetime. Unless they are leased, the customer must pay, or buy a new unit. So the sales-and-service approach could be quite profitable.

2

THE BASICS OF TELEPHONE EQUIPMENT TROUBLESHOOTING AND REPAIR

In this chapter we discuss the basic approaches for troubleshooting and repair of telephone equipment. We cover such areas as safety precautions, test equipment, tools, basic troubleshooting, and notes that apply to telephone equipment of all types. The troubleshooting procedures for specific types of telephone equipment are covered at the end of the related chapter.

Keep in mind that the information in this chapter is general in nature. If you are going to service a particular telephone, get all the service information you can on that instrument. Similarly, if you plan to go into telephone service on a large scale, study all the applicable service literature you can find. Then, when all else fails, you can follow instructions.

2-1. SAFETY PRECAUTIONS IN TELEPHONE EQUIPMENT SERVICE

In addition to a routine operating procedure (for both test equipment and the telephone), certain precautions must be observed during operation of any electronic test equipment. Many of these precautions are the same for all types of test equipment; others are unique to special test instruments, such as meters, oscilloscopes, and signal generators. Some of the precautions are designed to prevent damage to the test equipment or to the circuit where the service operation is being performed. Other precautions are to prevent injury to you. Where applicable, special safety precautions are included throughout the various chapters of this book.

2-1.1 General Safety Precautions

The following general safety precautions should be studied thoroughly and then compared to any specific precautions called for in the service literature and in the related chapters of this book.

Warning symbols on test equiptment. There are two standard international operator warning symbols found on some test equipment. One symbol, *a triangle with an exclamation point at the center*, advises the operator to refer to the operating manual before using a particular terminal or control. The other symbol, *a zigzag line simulating a lightning bolt*, warns the operator that there may be dangerously high voltage at a particular location, or that there is a voltage limitation to be considered when using a terminal or control. Always observe these warning symbols. Unfortunately, the use of symbols is not universal, particularly on older test equipment.

Metal cases and grounds. Many service instruments are housed in metal cases. These cases are connected to the ground of the internal circuit. In most cases, the grounded terminal of the instrument should be connected to the ground of the telephone being serviced. However, there are exceptions. For example, when a telephone product is plugged into the Model 1050 Telephone Analyzer described in Sec. 2-3, the ground lead of an oscilloscope or other earth-grounded test instruments should not be connected. Connect only the probe tip and leave the ground lead disconnected. Here's why.

As shown in Fig. 2-1, when a telephone is connected to the analyzer, one side of the telephone line is returned to earth ground through the analyzer. Because of the diode bridge (found in most telephone equipment), the telephone chassis (signal common) becomes a d-c potential with respect to this earth ground point. There is no particular disadvantage to this configuration. In fact, the analyzer provides exactly the same condition as a telephone plugged into a telephone company jack.

The ground leads of many test instruments are returned to ground through the a-c power cord. Since the analyzer already provides one earth point in the telephone, an earth-grounded test lead of another test instrument should not be connected to the telephone chassis (signal common). To do so *shorts one of the diodes in the bridge* as shown in Fig. 2-1. Because of the high impedance in series with the power supply from the tester, no damage occurs. However, the voltage supplied by the bridge drops to about 0.7 V, and the telephone circuits cease operation.

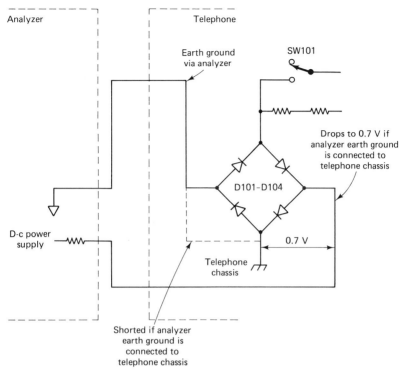

FIGURE 2-1. Ground connections between test analyzer and telephone circuits

Oscilloscope use. *The recommended procedure for oscilloscope use is to leave the ground lead disconnected and use only the probe tip.* A-c coupled measurements are not affected. For d-c coupled measurements, the probe tip may be initially touched to the signal common point and the vertical position control set to center the trace as reference. Subsequent measurements then represent the waveform or voltage with respect to the signal common point. If noise becomes a problem with the ground lead disconnected, it may be connected to the earth-ground point in the telephone. Both leads of most voltmeters are isolated from earth ground, and thus the signal common may be used for reference.

High voltages. Remember that there is always danger in servicing telephones that operate at hazardous voltages, especially if you pull off covers with the power cord connected. Fortunately, most telephone circuits operate at potentials well below the line voltage, since the circuits are essentially solid state. However, a line voltage of 120 V is sufficient to cause serious shock and possibly death! Always make some effort (such as reading the service literature) to familiarize yourself with the telephone before service, bearing in mind that line voltages may appear at unexpected points in a defective telephone.

Ring voltage and d-c (battery) operating power. The ring voltage is a 20-Hz square wave of at least 40 V, and possibly 80 to 130 V. This voltage can appear on a telephone line, or can be taken from a test instrument that simulates a telephone line (such as the analyzers described in Secs. 2-3 and 2-5). The ring voltage is applied to a telephone only in the on-hook condition (when the equipment is operating properly). If the ring voltage is applied to a telephone in the off-hook condition, even at the 40-V level, the internal circuits of a telephone can be destroyed. (The ring voltage can also destroy a service technician!)

The d-c (battery) operating power can also present a problem. This power starts as -48 V at the telephone company, but usually drops to between 5 and 12 V at the telephone. However, test instruments that simulate a telephone line produce the full 48 V at their outputs.

Keep in mind that an electrical shock causing 10 mA of current to pass through the heart will stop most human heartbeats, and voltages as low as 35 V should be considered dangerous (and possibly lethal).

Remove power. It is good practice to remove power before connecting test leads to high-voltage points. It is preferable to make all service connections with the power removed. Since this is generally impractical, be especially careful to avoid accidental contact with player circuits. Keep in mind that even low-voltage circuits may be a problem. For example, a screwdriver dropped across a 12-V line in a solid-state circuit can cause enough current to burn out a major portion of the telephone, possibly beyond repair. Of course, that problem is nothing compared to the possibility of injury to yourself! Working with one hand away from the telephone, and standing on a properly insulated floor, lessen the danger of electrical shock.

Capacitors may store a charge large enough to be hazardous, although generally not in solid-state circuits. Discharge capacitors before attaching test leads. (Make sure that you have turned off the power before you discharge the capacitors!)

Remember that leads with broken insulation offer the additional hazard of high voltages appearing at exposed points along the leads. Check test leads for frayed or broken insulation before working with them. To lessen the danger of accidental shock, disconnect test leads immediately after the test is complete.

Remember that the risk of severe shock is only one of the possible hazards. Even a minor shock, or touching a hot spot, can put you in danger of more serious risks, such as a bad fall or contact with a source of higher voltage.

The experienced service technician guards continually against injury and does not work on hazardous circuits unless another person is available to assist in case of accident, preferably someone with CPR training.

Even if you have considerable experience with test equipment used in service, always study the service literature of any instrument with which you are not thoroughly familiar.

Use only shielded leads and probes. Never allow your fingers to slip down to the metal probe tip when the probe is in contact with a hot or live circuit.

Avoid vibration and mechanical shock. Most electronic test equipment is delicate (more so than telephone equipment). So you should be particularly careful when handling test equipment, even the portable models. Also keep in mind that an electronic telephone (or answering machine, etc.) is more likely to be damaged by vibration and mechanical shock than is the standard telephone.

Study the circuit being serviced before making any test. Try to match the capabilities of the test instrument to the circuit being serviced. For example, if the circuit under test has a range of measurements to be made (ac, dc, RF, modulated signals, pulses, or complex waves), it is usually necessary to use more than one instrument. Most meters measure d-c and low-frequency signals. If an unmodulated RF carrier is to be measured, use an RF probe. If the carrier to be measured is modulated with low-frequency signals, a demodulator probe must be used. If pulses, square waves, or complex waves are to be measured, a peak-to-peak meter can possibly provide meaningful indications, but an oscilloscope is the logical instrument. If the problem is one of monitoring the digital logic pulses associated with the microprocessor found in some telephone equipment, you must use digital test equipment such as logic probes and pulses. Or you can try a really novel approach and use the test instrument recommended in the telephone equipment service literature! Note that the special test equipment described in Sec. 2-2 is designed specifically for telephone service, and includes the necessary signals and indicators.

Design alterations. Do not alter or add the mechanical or electrical design of telephone equipment. Design alterations, including (but not limited to) addition of auxiliary earphones, cables, accessories, and so on, might alter the safety characteristics of the telephone and create a hazard to the user (such as connecting one side of the power line to the user's ear). Any design alterations or additions may void the manufacturer's warranty and may make the *servicer responsible* for personal injury or property damage resulting therefrom.

Product safety notices. Many electrical and mechanical parts in telephones have special safety-related characteristics, some of which are often not evident from visual inspection, nor can the protection they give necessarily be obtained by replacing the parts with components rated for higher voltage, wattage, and so on. The manufacturers often identify such parts in their service literature. One common means of such identification is by *shading on the schematic and/or parts lists*, although all manufacturers do not use shading or limit such identification to shading. Always be on the alert for any special product safety notices, special parts identification, and so on. Use of a substitute part that does not have the *same safety characteristics* (not just the same electrical or mechanical characteristics) might create shock, fire, and/or other hazards. A simple way to solve the problem is to use the part recommended in the service literature!

Good electronic service practices. The author assumes that you are already familiar with good electronic service practices (removing the power cord before replacing circuit boards and modules, installing heat sinks as required on solid-state devices, etc.). The author also assumes that you can handle electrostatically sensitive (ES) devices (such as FETs, MOS chips, etc.); that you can solder and unsolder ICs, transistors, diodes, and so on; and that you can repair circuit-board copper foil as needed. If any of these seem unfamiliar to you, please, please do not attempt to service any telephone, especially the author's!

2-2. TEST EQUIPMENT FOR TELEPHONE SERVICE

Although some of the test equipment used in telephone service is the same as that for other fields of electronics (meters, signal generators, oscilloscopes, frequency counters, power supplies, and assorted clips, patch cords, etc.), you must also have certain test equipment that is unique to telephone service. For that reason, we go into considerable detail on special telephone test equipment in this book. Let us start with a general description of the special signals and indicator or monitor devices you need.

2-2.1 General Requirements

The following are minimum requirements for test signals and indicator/monitors used in telephone equipment service.

Telephone line simulator. Unless you want to tie up a telephone line (or possibly two lines for some tests) you must have some device to simulate a telephone line. A 48-V d-c power supply, fed through a 1.5-kΩ series resistance, will do for the basic telephone line simulator. Of course, the simulator must also have input/output jacks for applying ring voltage and test tones and for measuring dialing and audio signals. Keep in mind that if you use a real telephone line, and perform tests that produce shorts, abnormal voltages, and so on, on that line, you may hear a few choice words from the telephone company!

Ring-voltage generator. You must provide a 20-Hz, square-wave ring voltage. The ring voltage must be variable from about 40 to 100 V. Also, keep in mind that the ring voltage is applied to a telephone only in the on-hook condition (through capacitors to circuits that do not have their own operating power). If you apply the ring voltage to an off-hook telephone, you can destroy the other circuits. So your ring-voltage generator must have some scheme to shut off automatically if the telephone goes off-hook.

Dial decoder. You must have some means of monitoring the pulses (or tones) produced when the telephone is dialed, and that the correct number of pulses (or tone pairs) are produced for each digit.

RF level. If you are going to service cordless telephones, you must be able to measure the RF level of the transmitted signals. Of course, you can touch an RF probe (with meter or scope) to the telescope antenna of base unit and portable unit, but only when they are operating in the range 46 to 49 MHz. You will need a special tester to measure RF level of base units operating in the 1.7-MHz band (which use the a-c power cord as an antenna).

FM equipment. Since cordless telephones use FM, you need a *deviation meter* to measure transmitter deviation and an FM-modulated RF generator for receiver testing.

Frequency counter. A frequency counter used in telephone service must provide both the frequency range, and resolution, to measure transmitted RF carrier frequency and to set the RF signal generator on frequency. The frequency counter must also measure the guardtone and ring-signal frequencies, whether generated by a cordless telephone or by an audio-signal generator for injection into a cordless telephone. Again, the 1.7-MHz base unit transmitters can be a problem, since the radiated signal level (from the a-c power cord) may not be sufficient to get a good reading on some frequency counters.

One simple way to provide all of these special signals and monitoring devices is to use the test equipment described in the remainder of this chapter.

2-3. B&K-PRECISION MODEL 1050 TELEPHONE ANALYZER

Figure 2-2 shows the Model 1050 Telephone Analyzer. The 1050 is a full-featured test instrument for testing, troubleshooting, and adjusting corded telephones, cordless telephones, telephone answering machines, automatic dialers, and most other telephone products. The 1050 generates virtually all signals needed for telephone product servicing. These signals include d-c (battery) telephone voltage (-48 V), 20-Hz ring voltage, dial tone, ringback, fixed 1-kHz audio, variable 100-Hz to 10-kHz audio, and RF on all frequencies assigned for cordless telephone operation.

The 1050 also analyzes virtually all signals generated by telephone products. These include DTMF and pulse-dial varification, d-c levels, audio levels, relative RF level, RF frequency error, and FM deviation. Other important features include testing of telephone cords for short- or open-circuit paths, a variable 1.5- to 10-V d-c power source, and jacks to accommodate an oscilloscope or frequency counter for external measurements.

2-3.1 Corded Telephone Functions

The 1050 simulates a telephone exchange by generating a dial tone, ringing an on-hook telephone, and stops ringing when the telephone is off-hook. The 1050 also links two telephones together for voice-quality testing, and operates answer-

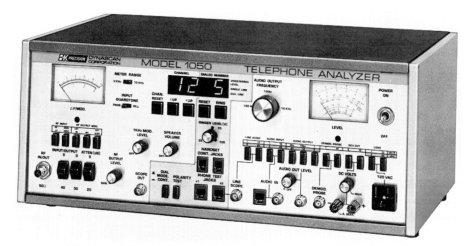

FIGURE 2-2. Model 1050 Telephone Analyzer (Courtesy of Dynascan Corporation)

ing machines, automatic dialers, and other telephone products. When the 1050 is used in the dial mode of operation, the telephone test jacks simulate the voltage and resistance of a telephone line. A dial tone identical to that provided by the telephone company is supplied at the test jacks, as well as ringing signals (jack 1 only), ringback tones (jack 2 only), and voice communication between the test jacks. A switch reverses the polarity of the d-c voltage applied to the first telephone test jack to allow testing of the automatic polarity circuits in the telephone.

When a telephone is plugged into jack 1, dialed numbers (from either a pulse- or tone-dial telephone) are decoded and displayed by the analyzer. For use with high-speed dialers, the numbers are slowly released in the same order as dialed.

During the ring test, ring voltage can be adjusted from 35 to 100 V (rms). This not only tests the ability of a telephone to ring, but also finds the threshold voltage of the telephone's ringer circuit. For intermittent ringing problems and signal tracing, the analyzer ring generator can be switched to a continuous ring cycle (in contrast to the normal 2 to $2^1/_2$ seconds on and $3^1/_2$ to 4 seconds off).

When the analyzer is switched to the continuity mode of operation, the telephone test jacks and handset continuity jacks are connected so that handset and telephone line cords can be tested for shorts and breaks. The polarity switch is used to verify that the cord is wired correctly.

2-3.2 Cordless Telephone Functions

The 1050 also includes many features for cordless telephone servicing. When used in the receive mode, the 1050 serves as a calibrated test receiver for servicing the companion base or portable transmitter. The 1050 shows whether or not a signal is being generated by the transmitter under test.

Step attenuators at the input of the 1050 permit relative RF power output measurement. A meter measures carrier frequency error and can be switched to measure transmitter modulation.

An audio output is available at an output terminal for connection to an oscilloscope, frequency counter, or other monitoring instrument. This permits convenient measurement of the guardtone or pilot frequency. The audio output is also applied to a speaker, permitting voice-quality and other listening tests.

When used in the transmit mode of operation, the 1050 serves as an RF signal generator, simulating the base or portable transmitter for servicing the companion receiver. A properly modulated RF output from the 1050 should result in an audible signal from the receiver under test (go/no-go test). Such a signal also permits signal tracing throughout the receiver under test. The RF output level is variable, permitting relative receiver sensitivity measurement. The RF output frequency is calibrated; thus a measurement in the receiver under test determines any carrier frequency error.

A built-in 1-kHz oscillator provides convenient internal modulation and a built-in VFO (variable-frequency oscillator) can be used to span the audio frequency and generate guardtones. Two external modulation jacks permit modulation by an audio generator or telephone (voice test). Internal and external modulation are independently selectable and adjustable, and both may be used simultaneously if desired.

The built-in frequency-error/modulation meter measures modulation or error, and the level meter measures a variety of a-c or d-c levels from input and output connectors on the front panel. The oscilloscope output jack permits measurement of the modulating signal. A frequency counter can be connected for accurately setting the internal audio generator to the desired guardtone or ring-signal frequency.

The 1050 operates on all radio frequencies allocated for cordless telephone use by the FCC, including the 1.7-, 46-, and 49-MHz bands. Digital channel selection is used. A PLL (phase-locked loop) and crystal control provide stable frequency precision, serving as a standard for checking and readjusting cordless telephone frequency-generating circuits.

The frequency-error/modulation meter has two ranges, 0–3 kHz and 0–10 kHz. The low range provides good resolution for carrier frequency-error measurement and for low modulation. The high range permits measurement of fully modulated signals. Also, when the range 0 to 3 kHz and RF error mode are selected, a filter is enabled that removes the effect of guardtone modulation. This allows frequency-error measurements to be made without disabling the guardtone signal in the equipment under test. The built-in 1-kHz oscillator output is also available at an output jack for external use. This is convenient for modulating the RF carrier of a cordless telephone transmitter under test.

The guardtone filter can be switched in or out, permitting measurement of guardtone modulation when desired, and rejecting guardtone and high-frequency noise (above 3 kHz) from the speaker for voice-listening tests.

2-3.3 Level Meter Functions

The level meter measures a variety of a-c and d-c signal levels. When the audio output mode is selected, three metering ranges are available. When any of the other modes are selected, two metering ranges are available. All a-c measurements can be made in either volts or dBm.

When the meter is used to measure telephone line audio, the meter shows the actual signal level that is present when operating the telephone on a telephone line. The meter may be used to measure both incoming and outgoing audio, as well as dialing tone signal level.

The meter may also be used to measure the level of signals fed to the audio input jacks. This is a convenient way to set externally generated signal levels for use as a modulation source when transmitting. When the external modulation level control is turned off, the audio input telephone jack is disabled and the input impedance of the BNC jack is switched from 600 Ω to 1 MΩ. This allows the audio input BNC to be used as a probe input, and the meter to be used as a general-purpose a-c voltmeter.

The meter may also be used to measure the level of a signal applied to the demodulator probe jack. This permits the use of a demodulator probe touched to an antenna of any cordless telephone to measure relative transmitter power.

When one of the d-c volts out pushbuttons is selected, the meter shows the d-c voltage that is available at the d-c voltage output jacks. This allows you to set the d-c voltage without the use of an external voltmeter.

The load function of the meter allows you to measure relative signal strength of 1.7-MHz-band RF transmitted from the base unit. This eliminates the need for special circuits to connect the RF signal to a signal-level meter.

2-3.4 Using the 1050 for Specific Tests

The remainder of this section is devoted to using the 1050 to perform specific tests on telephone equipment. The functions of the operating controls, as well as the operating procedures to perform specific tests, are described fully in the instruction manual for the 1050, and are not duplicated here. Instead, we concentrate on the *relationship* between the 1050 circuits, and the circuits within the telephone, during a particular test. This is done by showing both the 1050 test circuits (in block form), and the telephone circuits under test, for each test procedure.

A careful study of this relationship and the interconnections can help you to understand and troubleshoot the telephone circuits. If you know what is involved in testing a particular circuit, it often becomes quite clear what is wrong when that circuit does not perform properly. In any event, the test serves as a starting point for troubleshooting, as discussed in Sec. 2-7.

2-3.5 Operating Controls

Figure 2-2 shows the operting controls for the 1050. Reference to these controls is made throughout the remaining paragraphs of this section.

2-3.6 Cord Test

Figure 2-3 shows the test connections. This test is used to check a detachable handset cord (handset-to-desk unit) or detachable telephone cord (telephone-to-wall). *If the cord is not detachable at both ends, the cord cannot be tested as described here.*

If the telephone cord is found to be in proper working order, any problem can be isolated to the telephone or to the telephone exchange equipment. If the

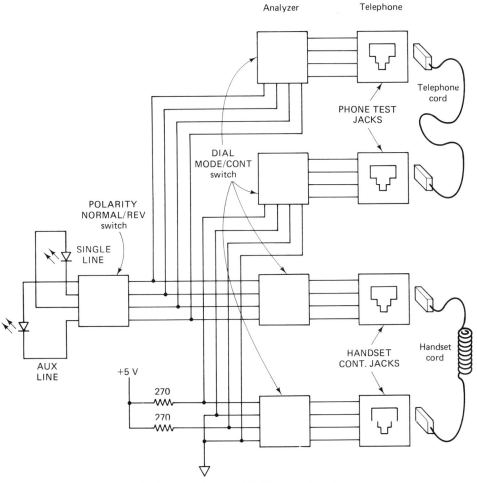

FIGURE 2-3. Model 1050 cord tests

cord is defective, other tests should still be carried out (after replacing the cord or cords) to ensure that there is no problem with the telephone itself.

For the cord test, both ends of a detachable telephone cord are plugged into the appropriate jacks. Two sets of jacks are provided, the PHONE TEST JACKS for checking detachable telephone cords (telephone-to-wall) and the HANDSET CONT. JACKS for detachable handset cords. The two sets of jacks are wired in parallel, and either type of cord is tested in the same manner, but only one may be tested at a given time.

When the DIAL MODE/CONT. button is engaged, a current is fed through each pair of wires (the two wires for voice, dialing, ringing, and power, as well as the two wires for power (for a lighted telephone). This current is taken from a +5-V supply (in the analyzer) through a 270-Ω resistor. At the other end of the cord, an LED is connected between one terminal of the cord jacks (one terminal and one LED for each pair of wires) and ground. One LED lights when a two-wire cord has continuity, with no shorts. Both LEDs light for a good four-wire cord. If either LED fails to light, the cord is defective, either shorted or no continuity.

When the POLARITY NORMAL/REV. switch is disengaged, the cord is tested as if wired properly. When the POLARITY NORMAL/REV. switch is engaged, the cord is tested as if the plugs on the cord are reversed.

Note that only one cord can be tested at a time, so do not plug in both cords simultaneously. Handset and telephone cords are generally not interchangeable and should not be plugged into the wrong jacks.

1. To test the telephone cord for continuity or shorts, plug both ends of the cord into the PHONE TEST JACKS and make sure that the DIAL MODE/CONT. button is engaged. If the SINGLE LINE indicator lights, the cord is good for normal telephone operation. If the AUX LINE indicator *also* lights, the cord is good for lighted telephone operation. The SINGLE LINE indicator shows the condition of the two wires in the cord used for dialing, ringing, conversation, and so on. The AUX LINE indicator shows the condition of the two wires that bring power to the night light, dial light, and so on. If either indicator fails to light, the cord is defective (unless it is a two-wire cord, which is rare).

For a proper test, gently bend and squeeze the cord to check for intermittent continuity or shorts. If the indicators go out, or flicker, the cord should be replaced.

2. Using the HANDSET CONT. JACKS, test the handset cord in the same way as the telephone cord.

Note that the POLARITY NORMAL/REV. button should be disengaged for this test. If either indictor fails to light, engage the POLARITY NORMAL/REV. button and check the indicators again. If the indicators light with the polarity reversed, the cord has reversed leads. However, unless the telephone is polarity sensitive, the phone should operate properly with this type of cord.

2-3.7 Dial Test

Figure 2-4 shows the test connections. This test applies to both tone-dial and pulse-dial (non-tone-dial) telephones. If tones are heard at the earpiece when digits are pressed, the telephone is a tone-dial model. If a series of clicks are

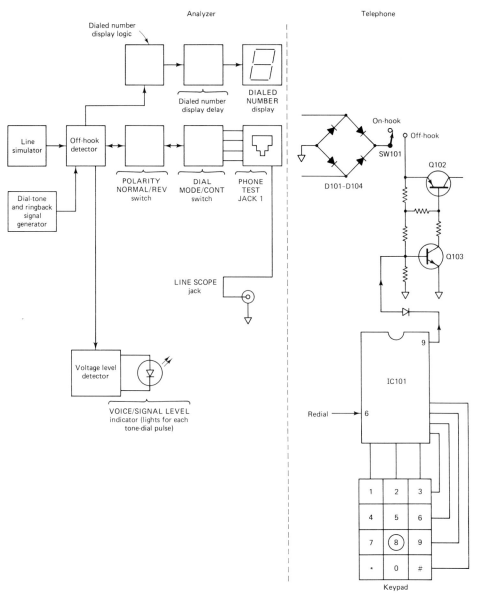

FIGURE 2-4. Model 1050 dial test

heard, the telephone is pulse-dial. When numbers are dialed faster than two digits per second, the analyzer stores up to 16 digits in memory, and then releases the digits at two digits per second. If you wish to clear the DIALED NUMBER display, press the RESET button. This clears the analyzer memory and places a "0" on the display (unless the telephone is still dialing).

For the dial test, the telephone is plugged into the PHONE TEST JACK 1. The *off-hook detector* circuit connects the *line simulator* circuit and the *dial tone and ringback signal generator* to the telephone through the POLARITY NORMAL/REV. switch; then the telephone is taken off-hook. The line simulator applies a negative d-c voltage through a 1.5-kΩ resistance. The POLARITY NORMAL/REV. switch reverses the d-c polarity to test the ability of the telephone to operate with either polarity. The dial tone and ringback signal generator generates two tones (440 and 352 Hz), mixes the tones together to produce a precision dial tone, and feeds this tone to the telephone.

As soon as the first digit is dialed from the telephone (from the keypad through IC101, Q102/Q103, SW101, and D101–D104) the dial tone is disconnected and the tones or pulses are fed to the *dialed number display logic* circuit, which decodes the pulses or tones into a 4-bit binary logic signal. The 4-bit binary code is then fed to the *dialed number delay*, which stores up to 16 dialed digits and then releases the digits to the DIALED NUMBER display (at approximately two digits per second so that the digits are easily read on the display).

The dialed number delay feeds the code to a decoder/driver which drives a seven-segment LED display. Each time a new digit is displayed, the DIALED NUMBER display is blanked momentarily.

1. Make sure that the DIAL MODE/CONT. button is disengaged. Plug the telephone to be tested into the PHONE TEST JACK 1. If a telephone is plugged into PHONE TEST JACK 2, make sure that the telephone is on-hook throughout this test.

2. Hang up the telephone under test (place the telephone on-hook).

3. Pick up the telephone under test (take the telephone off-hook) and listen for a dial tone. If no dial tone is present, this indicates that the telephone is not operating properly and should be repaired or replaced. (Trace the dial tone through the telephone audio circuits as described in Chapter 3.)

4. Press or dial each digit (be sure to use *all 10 digits*. With tone-dial telephones, it is also possible to test the "*" and "#" keys. The "*" key displays a decimal point (.) and the "#" key displays a bar (-) on the DIALED NUMBER display.

Each digit should appear on the DIALED NUMBER display in the same order as dialed. The display blinks momentarily each time a new digit is displayed so that if the same digit is dialed two or more times in sequence (for example, 933-3033), each individual digit can be distinguished. Also, each time a number is pressed on a tone-dial telephone, the VOICE/SIGNAL LEVEL indicator should light.

If the numbers do not appear on the DIALED NUMBER display in the correct order, or if the VOICE/SIGNAL LEVEL indicator does not light each time a digit is pressed on a tone-dial telephone, the telephone dialing circuits are not operating properly. (Check the telephone dial circuits as described in Chapter 3.)

If you wish to view the dialing pulses (from a pulse-dial telephone,) an oscilloscope can be connected to the LINE SCOPE jack on the front panel.

5. It is also possible to test the redial feature of a telephone. To do so, perform steps 1 through 4 of the dial test, and then operate the redial feature for the telephone being tested. Each digit should again appear on the DIALED NUMBER display in the same order as dialed. (If not, check the telephone redial circuits as described in Chapter 3.)

2-3.8 Ring Test

Figure 2-5 shows the test connections. This test applies to all types of telephones. *A word of caution before going on. Do not hold the telephone near your ear during the ring test. The ringing might be loud enough to cause hearing damage.*

For the ring test, the telephone to be tested is plugged into the PHONE TEST JACK 1. The *off-hook detector* circuit connects the telephone to the *line simulator, ring generator* and *ring amplifier* through the POLARITY NORMAL/REV. switch when the telephone is on-hook.

The ring generator produces a 20-Hz square-wave ring signal that is on for 2 to 2.5 seconds, and off for 3 to 3.5 seconds. The ring signal is passed through a low-pass filter which removes harmonics, and passes a 20-Hz sine wave. The signal is then fed through RINGER LEVEL (V) potentiometer, which sets the ring level, to the ring amplifier, where the signal is amplified to a level between 35 and 100 V (rms). This amplified ring signal is applied to the telephone ring circuits through C1 as shown in Fig. 2-5. Pulling out the RINGER LEVEL (V) controls causes the ring signal to be generated continuously rather than on and off. When the telephone is taken off-hook, the off-hook detector circuit inhibits the ring generator.

If a telephone is connected to PHONE TEST JACK 2 and taken off-hook during the ring test, the telephone receives the same signals.

1. Plug the telephone to be tested into PHONE TEST JACK 1, and hang up the telephone (place the telephone on-hook).

2. Adjust the RINGER LEVEL (V) control to 35, and press the RING button. Slowly increase the RINGER LEVEL (V) control setting until the telephone begins to ring. The telephone should ring until picked up (taken off-hook). When the telephone starts ringing, you have reached the threshold voltage of the telephone ringer circuit. If the telephone does not ring at all, even with the RINGER LEVEL (V) control turned to maximum, there is a problem with the ring circuits. (Check the ring circuits as described in Chapter 3.)

Typically, any telephone should ring with less than 100 V ringing voltage applied. (Most telephones will ring with as low as 40 V.) If not, the telephone

32 The Basics of Telephone Equipment Troubleshooting and Repair

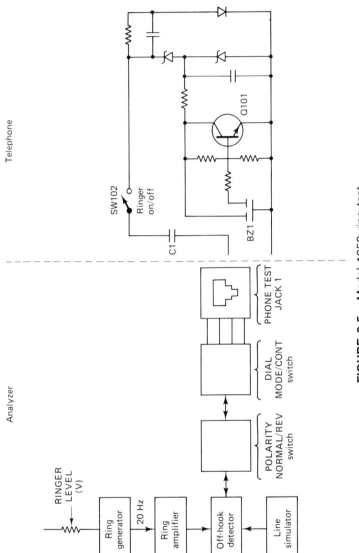

FIGURE 2-5. Model 1050 ring test

will not work with a long-line situation (when the telephone is to be used many miles from a switching station).

If you wish to hear the ringback tone while the telephone is ringing, listen to a second telephone connected to PHONE TEST JACK 2.

3. To cause the telephone to ring continuously (say, to trace the ring signal or measure the ring-signal level), pull the RINGER LEVEL (V) control out.

Note that if a large number of telephones or telephone-type devices are connected to a telephone line, this may prevent a good telephone from ringing. If the telephone being tested fails to ring when actually hooked up to a telephone line, but rings when connected to the analyzer, the line may be overloaded. Typically, a telephone line will ring five telephones or the equivalent. The *ringer equivalent number* (sometimes called R.E.N.) should be on the telephone or telephone-type device. If the total ringer equivalent (of all devices connected to one line) exceeds five, one or more of the telephones (or devices) may not ring, even though the telephone ring circuits are good.

2-3.9 Voice-Level Test

Figure 2-6 shows the test connections. This test applies to all types of telephones. The telephone to be tested is plugged into PHONE TEST JACK 1. The *off-hook detector* circuit connects the telephone to the *line simulator* and the *voltage-level detector* through the POLARITY NORMAL/REV. switch when the telephone is taken off-hook. The voltage-level detector rectifies the tone-dialing or voice signal into a full-wave voltage which feeds to an op-amp. The op-amp feeds the voltage to the VOICE/SIGNAL LEVEL indicator LED, with a 5-V reference connected to the other end of the LED. If the voltage applied to the LED is at a level sufficiently above 5 V, the LED lights.

1. Plug the telephone to be tested into PHONE TEST JACK 1 and pick up the telephone (take the telephone off-hook).
2. Press the RESET button to stop the dial tone. (It is important to turn off the dial tone while checking the VOICE/SIGNAL LEVEL indicator, since the dial tone causes the analyzer to read the voice level inaccurately.) The VOICE/SIGNAL LEVEL indicator should light, or flicker, when you talk into the telephone. This indicates that voice signals are being passed from the microphone through Q104, Q105, T101, Q102, SW101, and D101–D104 to the telephone line. If the indicator fails to light occasionally while you are talking, there can be a problem. (Check the audio circuits as described in Chapter 3.)
3. Set the POLARITY NORMAL/REV. button to the REV. position (engaged) and repeat steps 1 and 2 of the voice level test. This reverses the polarity of the power supplied to the telephone, and should not affect the operation of most telephones. Some telephones are not polarity guarded (typically,

34   The Basics of Telephone Equipment Troubleshooting and Repair

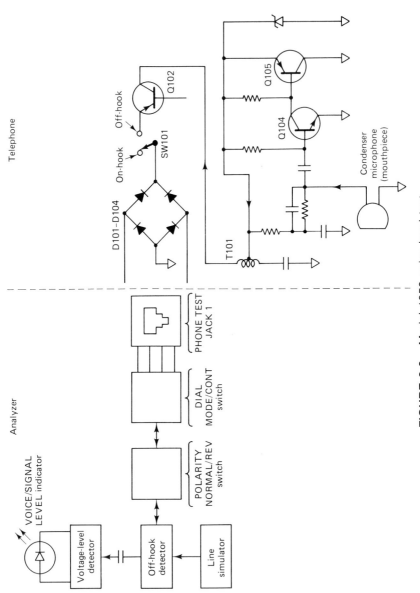

FIGURE 2-6. Model 1050 voice-level test

telephones that are at least several years old). Check the schematic diagram of the telephone being tested to see if there is polarity guarding. (Typically, polarity guarding is provided by diode rectifiers, such as the full-wave diode bridge (D101–D104) shown in Fig. 2-6.)

4. To check the level of the audio signal produced by voice, select the LINE AUDIO mode, and read the signal strength on the LEVEL meter.

2-3.10 Voice-Quality Test

Figure 2-7 shows the test connections. Two telephones of any type can be checked simultaneously. However, the test is most effective when one telephone is known to be in good working order (for both receiving and transmitting).

1. Plug the telephone to be tested into PHONE TEST JACK 1. Connect the second telephone (preferably a known good instrument) into PHONE TEST JACK 2.
2. Pick up both telephones (take them off-hook). You should now be able to talk and listen as if a telephone call has been completed between two telephones.

From telephone 1 you should be able to hear the person talking into tele-

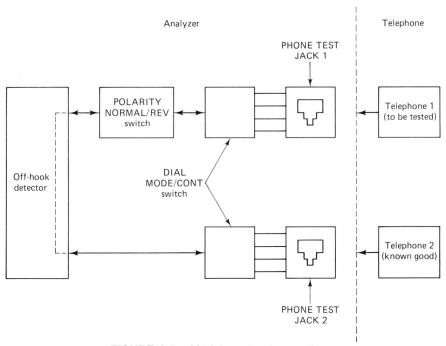

FIGURE 2-7. Model 1050 voice-quality test

phone 2 (the known good instrument). This checks the quality of the audio reception of telephone 1.

From telephone 2 you should be able to hear the person talking into telephone 1. This checks the quality of the audio transmission of telephone 1.

2-3.11 Cordless Telephone Tests (Non-RF)

To test the non-RF functions of a cordless telephone, plug the base-unit a-c power cord into the 120 VAC outlet, and plug the base-unit telephone cord into PHONE TEST JACK 1. With these test connections, the test procedures for cordless telephones are identical to the procedures for standard telephones, as described thus far in this section. However, be sure to follow the service literature for the telephone, as well as the information in Chapter 4. The analyzer can also be used to test the RF functions of a telephone as described next.

2-3.12 Cordless Telephone Tests (RF Functions)

The following paragraphs describe typical RF tests for cordless telephones. Although the descriptions apply to the 1050 analyzer, keep in mind that you must provide similar functions when troubleshooting cordless telephones if you do not have the 1050 (or an equivalent test instrument).

Attaching an antenna. The 1050 is provided with an *antenna coupling coil* for connecting a cordless telephone telescope antenna to the analyzer circuits, as shown in Fig. 2-8. The coil is used when checking 46/49-MHz cordless telephones. A connector-to-clip cable is recommended for 1.7/49-MHz telephones. Note that most cordless telephones use telescoping antennas for portable-to-base communication. However, a few compact cordless telephones use a *ferrite bar* in the portable unit, and the a-c line cord in the base unit.

Channel selection. The 1050 provides every frequency allocated for cordless telephones. Figure 2-9 shows the frequencies of the 99 channels. Always check the service manual to find the correct operating frequency of a specific cordless telephone. Usually, the channel number is indicated on the telephone. For example, if a Cobra cordless telephone is being tested and the channel indicated on the telephone is CH 1A, Fig. 2-9 (and the telephone service manual) shows that the base-to-portable frequency is 1.690 MHz, while the portable-to-base frequency is 49.830 MHz. Note that the information, such as CH 1A, shown in the COBRA CHANNEL column of Fig. 2-9, applies only to Cobra telephone equipment. However, the information in the CHANNEL and FREQUENCY columns is common to all cordless telephones used in the United States. The guardtone or pilot-signal frequency is also indicated on the telephone (such as G1, G2, etc.), but identification of the guardtone frequency is not necessary for channel selection (when using the 1050 analyzer). Of course, when you test cordless telephones

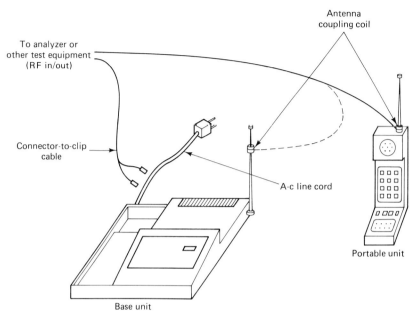

FIGURE 2-8. Connecting cordless telephone antennas to an analyzer or other test equipment for RF tests

without such an analyzer, you must also provide the guardtone signal, usually by means of a separate signal generator.

We will not go into the operation of the channel selection circuits, or the specific operating procedures for channel selection, in this book. The circuits and procedures are unique to the 1050. More important, the 1050 channel selection circuits are not connected directly to the circuits of the telephone under test, as is the case when the 1050 is used to check transmitted and received signals, which we discuss next.

Receiving signals. The 1050 can receive signals on any of the frequencies allocated for cordless telephones, both base unit and portable. Figure 2-10 shows the receive circuits in block form. The following is a brief description of the circuits.

RF input and routing. The RF signal (from the antenna coupling coil or connector-to-clip cable) is fed into the 1050 through the RF IN/OUT jack. The RF signal is fed through a *step attenuator*, where the signal can be attenuated from 0 to 90 dB, and then to a *signal direction control*, consisting of K201, Q208, and D211.

When none of the RF OUTPUT modes has been selected, K201 is deenergized and the incoming RF is fed to the *low-receive* and *high-receive* circuits. When any of the RF OUTPUT modes is selected, a logic high (5 V) is applied to Q208,

Channel	Frequency (MHz)	Cobra Channel
0	1.665	
1	1.665	
2	1.690	1A B-P
3	1.695	
4	1.710	7A B-P
5	1.725	
6	1.730	13A B-P
7	1.750	19A B-P
8	1.755	
9	1.770	25A B-P
10	46.610	
11	46.630	
12	46.670	
13	46.710	
14	46.730	
15	46.770	
16	46.830	
17	46.870	
18	46.930	
19	46.970	
20	49.670	
21	49.770	
22	49.830	1A P-B
23	49.845	7A P-B
24	49.860	13A P-B
25	49.875	19A P-B
26	49.890	25A P-B
27	49.930	
28	49.970	
29	49.990	
30, 32, 34, 36, 38	49.830	1A P-B
31, 33, 35, 37, 39	1.690	1A B-P

B-P, base-to-portable; P-B, portable-to-base.

FIGURE 2-9a. Cordless telephone channel frequencies (sheet 1 of 2)

which energizes K201, making a path for an outgoing signal and cutting off the incoming signals.

The *VCO select circuit* Q201/Q202 controls the *high/low receive input select* circuit (Q210, Q211, D207–D210). When the low VCO IC203 is enabled, the low-receive circuit is selected. The high-receive circuit is selected when the *high VCO* IC204 is turned on. Two VCOs (controlled by the channel selection circuits) produced outputs which are mixed with the incoming RF (from the telephone) to produce the correct IF for each channel.

The low-receive and high-receive circuits produce both a demodulated signal and an IF signal (of approximately 455 kHz). The *high/low receive output select* IC206 connects the demodulated audio signals to the *guardtone* filter IC25 and connects the IF signal to amplifiers Q4/Q5.

Channel	Frequency (MHz)	Cobra Channel
40, 42, 44, 46, 48	49.845	7A P-B
41, 43, 45, 47, 49	1.710	7A B-P
50, 52, 54, 56, 58	49.860	13A P-B
51, 53, 55, 57, 59	1.730	13A B-P
60, 62, 64, 66, 68	49.875	19A P-B
61, 63, 65, 67, 69	1.750	19A B-P
70, 72, 74, 76, 78	49.890	25A P-B
71, 73, 75, 77, 79	1.770	25A B-P
80	46.610	
81	49.670	
82	46.630	
83	49.845	
84	46.670	
85	49.860	
86	46.710	
87	49.770	
88	46.730	
89	49.875	
90	46.770	
91	49.830	
92	46.830	
93	49.890	
94	46.870	
95	49.930	
96	46.930	
97	49.990	
98	46.970	
99	49.970	

B-P, base-to-portable; P-B, portable-to-base.

FIGURE 2-9b. Cordless telephone channel frequencies (sheet 2 of 2)

Audio signal. When the INPUT GUARDTONE switch is in the REJ position, the demodulated audio signal is fed through IC25, which is a low-pass filter with a break frequency at 3 kHz, and a rolloff rate of about 20 dB/octave. From IC25, the signal is fed to buffer IC17B. When the INPUT GUARDTONE switch is in the PASS position, the filter is bypassed and the signal is fed directly to IC17B.

The audio signal from IC17B is fed to the loudspeaker through the SPEAKER VOLUME control and amplifier IC22. The audio is also fed to the SCOPE OUT jack and *AC/DC converter* IC20/21. The converter changes the a-c signal to a d-c voltage, which is fed to the Δ F/MOD meter M1 through *buffer* IC18B, when MOD(DEV) is selected.

IF signal. The IF signal from IC206 is fed to *mixer* IC23 through amplifiers Q4/Q5. Mixer IC23 also receives a 455-kHz signal from 4.55-MHz oscillator X1/Q2 and divide-by-10 circuit IC26. The output of IC23 is the difference

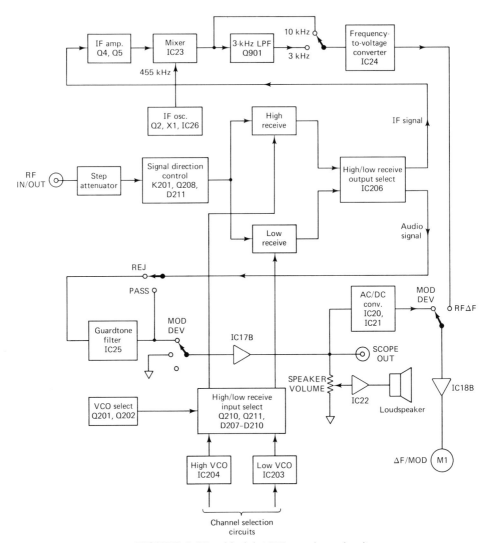

FIGURE 2-10. Model 1050 receiver circuits

between the two signals. When the telephone carrier is on-frequency, the resulting IF is 455 kHz and the mixer IC23 output (reference minus IF) is zero (or very close to zero).

When the 3-kHz range is selected (by the METER RANGE switch), the output of IC23 is fed to frequency-to-voltage converter IC24 through low-pass filter Q901, which reduces errors due to modulation. When the 10-kHz range is selected, the output of IC23 is fed directly to IC24. In both ranges, IC24 produces a d-c voltage proportional to *carrier frequency error*. This d-c voltage is fed to ΔF/MOD meter M1 through IC18B.

Checking portable-unit carrier frequency error. Figure 2-11 shows the test connections.

1. Set the 1050 to the channel that corresponds to the portable-to-base frequency of the cordless telephone (Fig. 2-9).
2. With the arrow (printed on the coupling coil label) pointing toward the

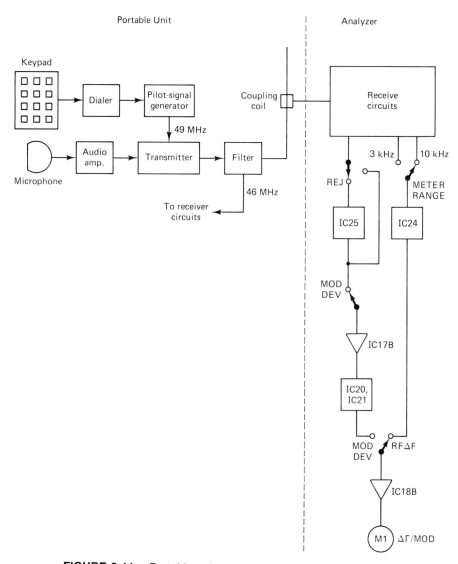

FIGURE 2-11. Portable-unit carrier frequency-error test connections

base of the antenna, place the coupling coil over the fully extended antenna. Hold the portable unit as during normal operation throughout the test. If the portable unit has no external antenna, use a connector-to-clip cable by holding the clips near the cordless telephone ferrite antenna.

3. Set the METER RANGE switch to 10 kHz, the INPUT GUARDTONE switch to REJ, and press the RFΔF button. The ΔF/MOD meter now shows the amount of offset (error) from the carrier frequency (indicated in kilohertz on the upper scale). If the reading is above 3 kHz, adjust the carrier frequency of the portable unit for as low an error reading as possible. If the reading is below 3 kHz (initially or after adjusting), set the METER RANGE switch to 3 kHz. Now the deviation is read on the lower scale (but still in kilohertz).

If the reading is above 2 kHz, the portable unit carrier frequency is off by a considerable amount. This may indicate that it is necessary to troubleshoot the portable transmitter circuits, in addition to adjustment of the circuits.

Note that the ΔF/MOD meter does not indicate whether the frequency is off in a positive or negative direction. Watch the meter closely as the carrier frequency is adjusted. If the meter reading increases as you adjust frequency, the adjustment is in the wrong direction. Always adjust the portable transmitter frequency to get a minimum reading on the ΔF/MOD meter.

It is generally not necessary to adjust the carrier frequency of a portable unit if the initial reading is less than 2 kHz. However, always consult the telephone service manual for recommendations.

When using the 1050, begin testing carrier frequency error with the METER RANGE switch in the 10-kHz position. In the 3-kHz position, the low-pass filter is used to remove the effects of guardtone modulation. The filter not only removes modulation above 3 kHz, but also the frequency error greater than 3 kHz. So there may be very low meter readings on the 3-kHz range when the error is much greater than 3 kHz.

Note that it is possible to check and adjust the portable-unit carrier frequency using conventional test equipment, although the procedures are more complex. Chapter 5 describes some typical procedures.

Checking guardtone or pilot signal. Figure 2-12 shows the test connections.

1. Set the 1050 to the channel that corresponds to the portable-to-base frequency of the cordless telephone (Fig. 2-9).
2. With the arrow (printed on the coil) pointing toward the base of the antenna, place the coupling coil over the fully extended antenna, or use a connector-to-clip cable as shown.

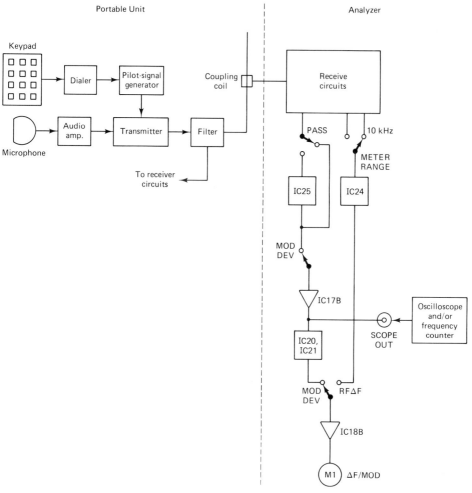

FIGURE 2-12. Portable-unit guardtone or pilot-signal test connections

3. Check that the carrier frequency is within 2 kHz (or as specified in the telephone service manual) before checking the guardtone. Then set the METER RANGE switch to 10 kHz and press the MOD(DEV) button. The ΔF/MOD meter now shows the deviation (indicated in kilohertz on the upper scale) caused by the modulating signal.

4. Set the INPUT GUARDTONE switch to PASS. This passes the guardtone or pilot signal and other signals above 3 kHz. The guardtone is an audio signal used to lock the base unit onto the signal being sent by the portable unit. (Typical guardtone or pilot-signal frequencies are 5.2 or 5.3 kHz.)

5. Connect an oscilloscope or frequency counter to the SCOPE OUT jack as

shown in Fig. 2-12. If you use the oscilloscope, a vertical setting of 0.5 V/div and a horizontal setting of 0.2 ms/div are recommended. Check that the frequency of the output matches the guardtone or pilot signal frequency specified in the telephone service manual. If the frequency is not within specifications, adjust the frequency before going on with other tests.

6. To test the *relative transmitter power* of the portable unit, transmit a signal to the 1050 from a *known-good* cordless telephone portable unit, and note how much attenuation you can select and still receive the signal clearly. The portable unit under test should be able to transmit a signal to the 1050 with approximately the same amount of attenuation and still produce a clear signal.

If the portable unit has a *mute switch*, keep the switch pressed when checking guardtone. If there is no mute switch, cover the mouthpiece and be careful not to allow any sound to reach the mouthpiece. If any audio other than the guardtone is transmitted by the portable unit, the audio causes the guardtone frequency reading to be inaccurate.

Note that it is possible to check and adjust the portable-unit guardtone or pilot-signal frequency using conventional test equipment, although the procedures are more complex. Chapter 5 describes some typical procedures.

Checking portable-unit modulation and audio quality. Figure 2-13 shows the test connections.

1. Set the 1050 to the channel that corresponds to the portable-to-base frequency of the cordless telephone (Fig. 2-9).
2. With the arrow (printed on the coil) pointing toward the base of the antenna, place the coupling coil over the fully extended antenna, or use a connector-to-clip cable as shown.
3. Set the INPUT GUARDTONE switch to REJ. This eliminates the guardtone or pilot signal and all high-frequency (above 3 kHz) noise.
4. Set the METER RANGE switch to 10 kHz. While speaking into the portable-unit mouthpiece, watch the ΔF/MOD meter. The meter needle should deflect away from zero (typical readings are about 1 to 2 kHz). This indicates the *carrier deviation* caused by the modulating signal. Deflection should increase as you speak more loudly. If the meter reading remains at or below 3 kHz, set METER RANGE to 3 kHz for more accurate readings.
5. While still speaking into the portable-unit mouthpiece, adjust the SPEAKER VOLUME control until the audio output at the speaker reaches a suitable level. The audio signal from the speaker should be clear as you speak. If desired, connect an oscilloscope to the SCOPE OUT jack to view the audio.
6. To monitor the guardtone or pilot signal, or any other high-frequency audio, set the INPUT GUARDTONE switch to PASS. Although some cordless telephones have guardtones above 3 kHz, the typical telephone company frequency range is between 300 and 3000 Hz. So, even though you can hear the guardtone

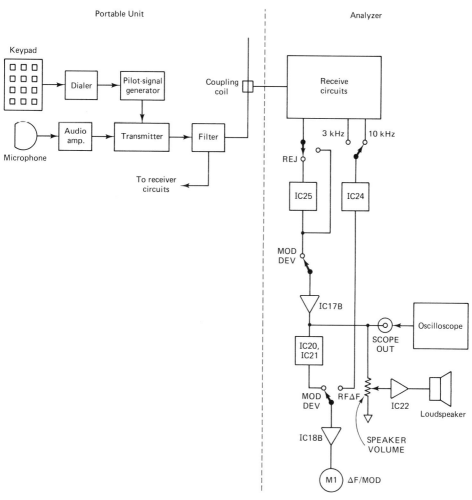

FIGURE 2-13. Portable-unit modulation and audio-quality test connections

audio through the speaker, the audio is not heard at the other end of the telephone line.

Checking base-unit carrier frequency error. Figure 2-14 shows the test connections.

1. Disengage the DIAL MODE/CONT. button. Plug the base-unit power cord into the 120 VAC outlet, and plug the base-unit cord into PHONE TEST JACK 1.
2. Set the 1050 to the base-to-portable frequency of the cordless telephone, either 1.7 or 46 MHz.

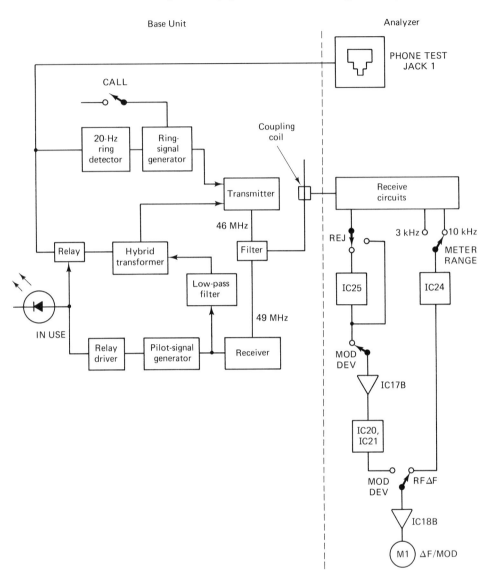

FIGURE 2-14. Base-unit carrier frequency-error test connections

3. If the 46-MHz band is used, place the coupling coil over the fully extended antenna. If the 1.7-MHz band is used, use a connector-to-clip cable by holding the clips near the base-unit a-c power cord.

4. Put the portable unit into the talk mode and press the RESET button (on the 1050) to cancel the dial tone. This causes the base unit to generate an unmodulated carrier signal. (To measure frequency error, or ΔF, always use an

unmodulated carrier. Any modulation (dial tone, ring signal, audio tone, etc.) causes an error in the reading.

If the portable unit is not working properly, there are several ways to make the base unit transmit an unmodulated RF carrier signal. For example, pressing the CALL button on the base unit produces a modulated carrier. However, immediately after the CALL button is released, the base unit continues to transmit an unmodulated carrier before the transmitter turns off. Measure ΔF during this delay period. Here are some other ways to produce unmodulated RF from the base unit.

If the base unit has no CALL BUTTON, plug the base unit into PHONE TEST JACK 1 and press the RING button. Measure the carrier frequency *between ring cycles* (when the carrier is unmodulated).

When a base unit is first plugged into the 120 VAC jack, an unmodulated RF carrier is generated for a brief period. This period can be used to measure frequency error. The period during which an unmodulated RF carrier is generated varies from one cordless telephone to another. If the period is too short for the meter to settle, you may have to open the base unit and trip the relay.

5. Set the METER RANGE switch to 10 kHz, and press the RFΔF button. Under these conditions, the ΔF/MOD meter shows the amount of offset or error from the carrier frequency (indicated in kilohertz on the upper scale). If the reading is above 3 kHz, adjust the base-unit carrier frequency for as low a reading as possible. If the reading is below 3 kHz (initially or after adjustment), set the METER RANGE switch to 3 kHz. Now the deviation is read on the lower scale (but still in kilohertz).

If the reading is above 2 kHz, the base-unit carrier frequency is off by a considerable amount. This may indicate the need to troubleshoot the base transmitter circuits, in addition to adjustment of the circuits.

Note that the ΔF/MOD meter does not indicate whether the frequency is off in a positive or negative direction. Watch the meter closely as the carrier frequency is adjusted. If the meter reading increases as you adjust frequency, the adjustment is in the wrong direction. Always adjust the base transmitter frequency to get a minimum reading on the ΔF/MOD meter.

It is generally not necessary to adjust the carrier frequency of a base unit if the initial reading is less than 2 kHz. However, always consult the base-unit service manual for recommendations.

When using the 1050, begin testing carrier frequency error with the METER RANGE switch in the 10-kHz position. In the 3-kHz position, the low-pass filter is used to remove the effects of modulation. The filter not only removes modulation above 3 kHz, but also any frequency error greater than 3 kHz. So there may be very low meter readings on the 3-kHz range when the error is much greater than 3 kHz.

Note that it is possible to check and adjust the base-unit carrier frequency using conventional test equipment, although the procedures are more complex. Chapter 5 describes some typical procedures.

Checking ring signal. Figure 2-15 shows the test connections.

1. Set the 1050 to the base-to-portable frequency of the cordless telephone, either 1.7 or 46 MHz.

2. If the 46-MHz band is used, place the coupling coil over the fully extended antenna. If the 1.7-MHz band is used, use a connector-to-clip cable by holding the clips near the base-unit a-c power cord.

3. Disengage the DIAL MODE/CONT. button. Plug the base-unit power cord into the 120 VAC outlet, and plug the base-unit cord into the PHONE TEST JACK 1. Set up the base unit (and portable unit) as if you are waiting for a telephone call (so that the base unit is on-hook).

4. Set the INPUT GUARDTONE switch to REJ. This eliminates all high-frequency (above 3 kHz) noise. No guardtone or pilot signal is generated by the

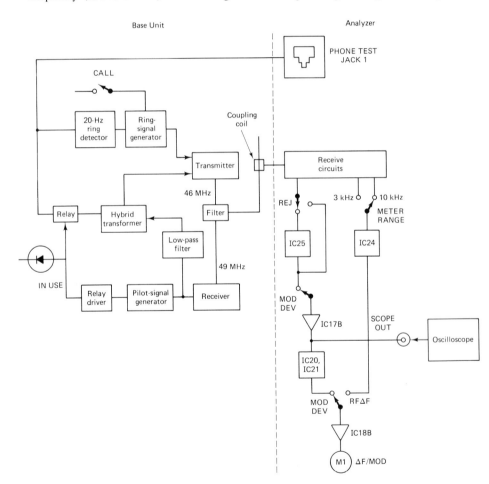

FIGURE 2-15. Base-unit ring-signal test connections

base unit. However, there still might be some high-frequency noise that you wish to eliminate.

5. Set the RINGER LEVEL (V) control to 100, and press the RING button. Pull out the RINGER LEVEL (V) control to cause *continuous ringing*. Connect an oscilloscope to the SCOPE OUT jack. (You can use a frequency counter for this test, but the counter will not tell you the amplitude of the ring signal.) If you use an oscilloscope, a vertical setting of 0.5 V/div and a horizontal setting of 0.2 ms/div are recommended.

6. Check that the frequency and amplitude of the base-to-portable ring signal match the specifications in the telephone service manual. If the frequency is not within specifications, adjust the frequency as necessary. If the amplitude of the ring signal is low, troubleshooting of the base-to-portable ring circuits may be indicated.

Checking base-unit modulation and audio quality. Figure 2-16 shows the test connections.

1. Set the 1050 to the channel that corresponds to the base-to-portable frequency (Fig. 2-9).

2. With the arrow (printed on the coil) pointing toward the base of the antenna, place the coupling coil over the fully extended antenna or use a connector-to-clip cable as shown.

3. Disengage the DIAL MODE/CONT. button. Plug the base-unit a-c power cord into the 120 VAC outlet, and plug the base-unit telephone cord into PHONE TEST JACK 1. Plug a standard corded telephone into PHONE TEST JACK 2, and take the standard telephone off-hook. This cancels the dial tone when the base unit goes off-hook.

4. Set the INPUT GUARDTONE switch to REJ. This eliminates the guardtone or pilot signal and all high-frequency (above 3 kHz) noise.

5. Set the METER RANGE switch to 10 kHz. While speaking into the *corded telephone* mouthpiece, watch the ΔF/MOD meter. The meter needle should deflect away from zero (typical readings are about 1 to 2 kHz). This indicates the *carrier deviation* caused by the modulating signal. Deflection should increase as you speak more loudly. If the meter reading remains at or below 3 kHz, set METER RANGE to 3 kHz for more accurate readings.

6. While still speaking into the corded telephone mouthpiece, adjust the SPEAKER VOLUME control until the audio output at the speaker reaches a suitable level. The aduio signal from the speaker should be clear as you speak. If desired, connect an oscilloscope to the SCOPE OUT jack to view the audio.

7. To test the *relative transmitter power* of a base unit (where the base-to-portable frequency is in the 46-MHz band), use the connections shown in Fig. 2-16, and transmit a signal to the 1050 from a *known-good* cordless telephone base unit, note how much attenuation you can select and still receive the signal clearly. The base unit under test should be able to transmit a signal to the 1050 with

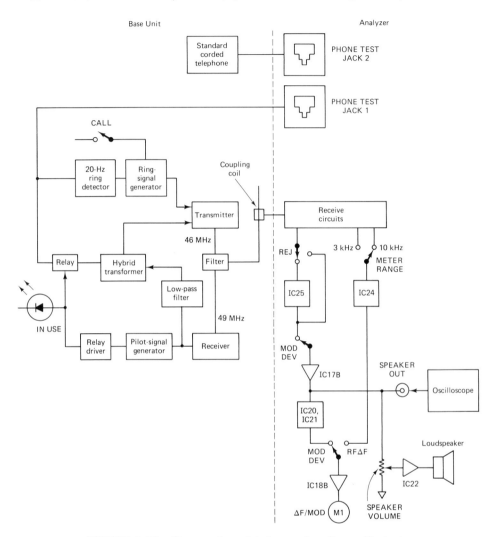

FIGURE 2-16. Base-unit modulation and audio-quality test connections

approximately the same amount of attenuation and still produce a clear signal.

8. To test the *relative transmitter power* of a base unit (where the base-to-portable frequency is in the 1.7-MHz band), press one of the two LOAD pushbuttons and read the relative transmitter power on the LEVEL meter. This is possible because base units that operate in the 1.7-MHz band transmit the RF signal over the a-c power line. Figure 2-17 shows the equivalent load that terminates the base-unit transmitter. Note that the readings are taken across a 50-Ω resistance. Although this does not represent the total RF signal (the total load, including the capacitive reactance, may be about 150 Ω at 1.7 MHz), it produces a reading

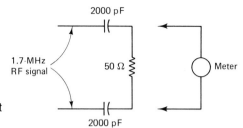

FIGURE 2-17. Equivalent load that terminates the base-unit transmitter

relative to the total RF signal. A known-good base unit may be used to establish a reference load if desired. In any case, most base unit transmitters produce a reading of about 3.5 to 4 V when the 10 V range is selected on the LEVEL meter.

Note that due to the inherent characteristics of demodulator circuits, as signal levels go below about 0.7 V (rms), the voltage used to power the diodes becomes a significant part of the input voltage. This causes the output reading on the LEVEL meter to become inaccurate as input voltages drop below 0.7 V.

Transmitting signals. The 1050 can transmit signals on any of the frequencies allocated for cordless telephones, both base unit and portable, and provides a modular telephone jack and a BNC connector for externally generated modulating signals. Also provided are an internally generated 1-kHz sine wave and an internally generated variable-frequency (100 Hz to 10 kHz) sine wave. External or internal (or both simultaneously) modulating signals are selectable using the RF OUTPUT controls. Both the external and internal signal levels are adjustable using the EXT MOD LEVEL, 1-kHz MOD LEVEL, and AUDIO OUT LEVEL controls. When using the 1050 as a transmitter, it is best to keep the SPEAKER VOLUME turned OFF unless you wish to listen to the signal being used to modulate the carrier signal.

Figure 2-18 shows the transmit circuits in block form. The following is a brief description of the circuits.

RF path. From the VCO (high-VCO or low-VCO, whichever one is enabled), the RF signal is fed to the variable attenuator Q203, where the amount of attenuation is set by the front-panel RF LEVEL control. From Q203, the signal is fed to the signal direction control circuit (Q208, D211, K201) through output buffer Q204–Q207. From the signal direction control circuit, the signal is routed to the front-panel RF IN/OUT jack through the step attenuator.

Modulation. There are three choices of modulating signals, all controlled by the three RF OUTPUT controls (INT, EXT, EXT & INT). The 1-kHz oscillator (IC116, Q1) feeds a 1-kHz signal to the RF OUTPUT switch unit through the 1-kHz MOD LEVEL control (S9, VR3). The 1-kHz MOD LEVEL control adjusts the level, or disconnects (when the control is fully counterclockwise) the 1-kHz signal.

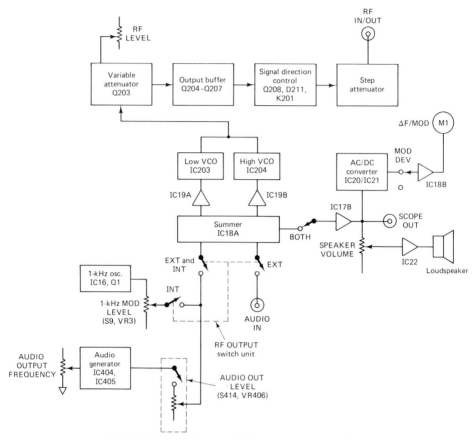

FIGURE 2-18. Model 1050 transmitter circuits

The variable-frequency audio oscillator IC405 feeds an audio signal (adjustable from one hundred to ten thousand Hz) to amplifier IC404 through the AUDIO LEVEL control (S414, VR406). The AUDIO LEVEL control adjusts the level, or disconnects (when the control is fully counterclockwise) the audio signal. The amplified audio signal is fed to the AUDIO OUT jack, the AC/DC converter, and the RF OUTPUT switch unit.

The externally generated signals are also fed to summer IC18A through the RF OUTPUT switch unit. In the INT mode, the internally generated signals (only) are fed to the summer. Externally generated signals (only) are fed in the EXT mode. Both internal and external signals are fed in the EXT & INT mode.

The summer IC18A is used to mix the internal and external signals (when both are used) and to act as a buffer when either internal or external signals are used. Signals from IC18A are sent to three buffers. One buffer IC19A feeds the signal to the low-VCO where the signal is used to modulate the low-band carrier. Buffer IC19B feeds the signal to modulate the high-VCO.

B & K-Precision Model 1050 Telephone Analyzer   53

Monitoring. Buffer IC17B feeds the signal to the AC/DC converter IC20/IC21, the SCOPE OUT jack, and to the monitor speaker through SPEAKER VOLUME control and amplifier IC22. The SCOPE OUT jack is used to drive an oscilloscope or frequency counter for monitoring the audio. The AC/DC converter changes the audio into a corresponding d-c voltage which is applied to ΔF/MOD meter M1 through buffer IC18A.

Checking portable-unit reception. Figure 2-19 shows the test connections.
1. Set the 1050 to the channel that corresponds to the base-to-portable frequency (Fig. 2-9).

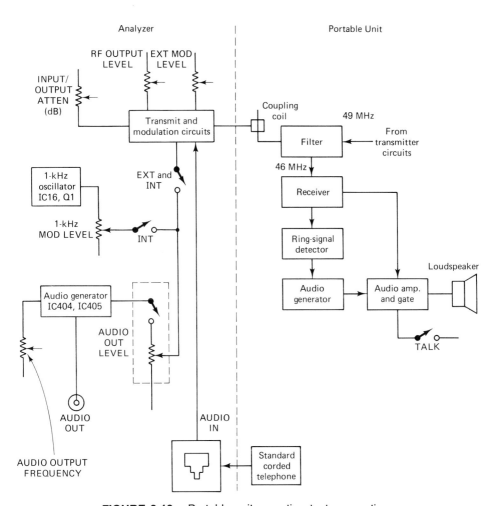

FIGURE 2-19. Portable-unit reception test connections

2. With the arrow (printed on the coil) pointing toward the base of the antenna, place the coupling coil over the fully extended antenna (46-MHz band), or hold the coil near the ferrite antenna as shown (1.7-MHz band).

3. Set the RF OUTPUT LEVEL and EXT MOD LEVEL controls to the MIN position (fully counterclockwise). Turn on all of the attenuators (slide them down).

4. Select the INT mode, and put the portable unit in the talk mode. Turn the AUDIO OUT LEVEL control to OFF, and adjust the 1-kHz MOD LEVEL, RF OUTPUT LEVEL, and INPUT/OUTPUT ATTEN (dB) controls until the 1-kHz tone is heard in the portable-unit speaker. If the tone cannot be heard with the 1-kHz MOD LEVEL and RF OUTPUT LEVEL controls turned all the way up and all INPUT/OUTPUT ATTEN (dB)controls set to 0, the portable unit is not receiving the signal properly, and must be repaired.

Note that the amount of attenuation you can select and still receive the signal is an indication of the portable-unit receiver-circuit sensitivity. Most portable units should be able to receive the signal with about 80 dB of attenuation (70 dB on the step attenuators and the RF OUTPUT control set to midrange). However, the 80-dB figure should be used as a guideline. (All cordless telephone portable units are not equal, as you will soon find out!)

5. Once it is determined that the portable unit is receiving properly, turn the 1-kHz MOD LEVEL control to OFF and turn the AUDIO OUT LEVEL control on. Connect an oscilloscope or frequency counter to the AUDIO OUT jack. Set the internal generator to produce a 0.5-V (rms) 300-Hz sine wave using the AUDIO OUTPUT FREQUENCY and AUDIO OUT LEVEL controls. (The voltage can be set using the LEVEL meter indication.) It should not be necessary to adjust the RF OUTPUT or INPUT/OUTPUT ATTEN (dB) controls (from the settings established in step 4) to get sufficient volume at the portable unit. If sufficient volume cannot be obtained, the audio bandwidth of the portable unit is not within specifications. (There is a drop-off at the low end.)

6. Repeat step 5 using a 0.5-V (rms) 3000-Hz signal in place of the 300-Hz signal. This checks the frequency response over the full range of a typical portable unit. (The frequency response is checked at 1000 Hz and 300 Hz in steps 4 and 5, respectively.) If the response is good at these three frequencies, it is reasonable to assume that the response is good across the entire range.

7. Plug a corded telephone into the AUDIO IN modular telephone jack. Set the 1050 to the EXT RF OUTPUT mode. Adjust the EXT MOD LEVEL control while speaking into the corded telephone.You should be able to hear what is being said into the corded telephone (clearly) from the portable-unit speaker (earpiece). Again, the RF OUTPUT and INPUT/OUTPUT ATTEN (dB) controls should not need adjustment from the settings established in step 4.

Checking base-unit reception. Figure 2-20 shows the test connections.

1. Set the 1050 to the channel that corresponds to the portable-to-base frequency (Fig. 2-9).

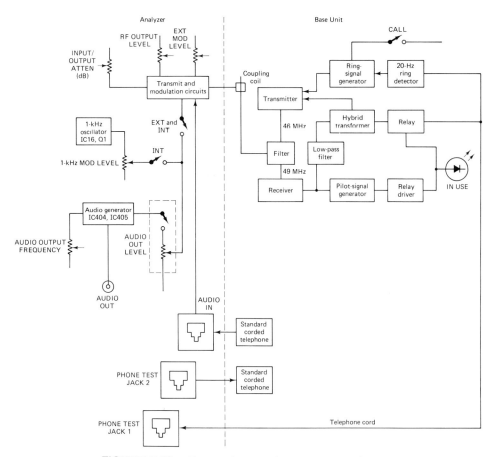

FIGURE 2-20. Base-unit reception test connections

2. With the arrow (printed on the coil) pointing toward the base of the antenna, place the coupling coil over the fully extended antenna as shown.

3. Set the RF OUTPUT LEVEL and EXT MOD LEVEL controls to the MIN position (fully counterclockwise). Turn on all of the attenuators (slide them down).

4. Plug the base-unit a-c power cord into the 120 VAC connector, and plug the base-unit telephone cord into PHONE TEST JACK 1. Plug a corded telephone into PHONE TEST JACK 2 and take the phone off-hook. Disengage the DIAL MODE/CONT. button throughout the test. Select the INT mode of RF OUTPUT, and turn the 1-kHz MOD LEVEL control off.

5. Find the guardtone or pilot-signal frequency of the cordless telephone under test in the service literature.

6. Set the 1050 internal signal generator to the guardtone frequency (using a frequency counter connected to the AUDIO OUT jack). (Typical guardtone signals for the Cobra cordless telephones are 5.3, 6.0, or 6.7 kHz.) Set the output level

of the guardtone signal to 0.5 V (rms) using the AUDIO OUT LEVEL control, AUDIO OUTPUT buttons, and LEVEL meter.

7. Adjust the RF OUTPUT LEVEL and INPUT/OUTPUT ATTEN (dB) controls until the relay in the cordless telephone base unit energizes. (Some base units have an IN USE light that goes on when the relay is energized.) The base unit should now be in the talk (off-hook) mode.

8. Adjust the 1 kHz MOD LEVEL control until the 1-kHz signal is heard at the earpiece of the corded telephone plugged into PHONE TEST JACK 2. Most base units should be able to receive the signal when about 80 dB of attenuation is selected (the step attenuators set to 70, and the RF OUTPUT LEVEL control set midrange). For *relative receiver sensitivity measurements*, check a *known-good* base unit to see how much attenuation can be selected, while still receiving the signal. Use those settings as a reference.

9. Plug a corded telephone into the AUDIO IN modular telephone jack. Keep all other connections and controls undisturbed. Turn off the 1-kHz MOD LEVEL (fully counterclockwise). You should be able to hear (clearly) what is being said into the telephone (at the AUDIO IN jack) from the earpiece of the other telephone (connected at the PHONE TEST JACK 2).

2-4. B&K-PRECISION MODEL 1047 CORDLESS TELEPHONE TESTER

Figure 2-21 shows the Model 1047 Cordless Telephone Tester. The 1047 checks all RF functions of cordless telephones in essentially the same way as the Model 1050 described in Sec. 2-3. For that reason, we do not duplicate these functions

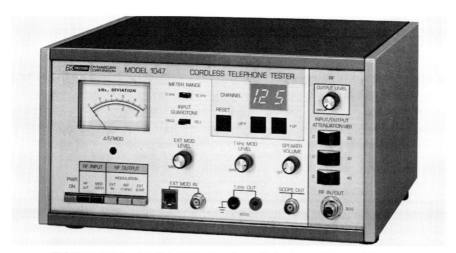

FIGURE 2-21. Model 1047 Cordless Telephone Tester (Courtesy of Dynascan Corporation)

here. However, the 1047 does not check the telephone line interface circuits in the base unit, such as verification of 20-Hz ring voltage detection from the telephone line, verification of correct tone or pulse dialing to the telephone line, and verification of voice interface circuits to the telephone line.

The manufacturer recommends that after verifying proper operation of all RF functions with the 1047, the non-RF functions should be checked by connecting the base unit to a telephone line, followed by placing and receiving a call.

A more convenient alternative is to use a B&K-Precision Model 1045 Telephone Product Tester, described in Sec. 2-5. This avoids tying up a telephone line, fully simulates an operating telephone line, and permits testing and troubleshooting of defective non-RF circuits.

If all RF and non-RF functions check out properly but there is still a problem (the telephone will not operate on a particular line), the B&K-Precision Model 1042 Telephone Line Analyzer can be used to test the telephone line. The 1042 is described in Sec. 2-6.

2-5. B&K-PRECISION MODEL 1045 TELEPHONE PRODUCT TESTER

Figure 2-22 shows the Model 1045 Telephone Product Tester. The 1045 performs a wide variety of go/no-go tests on standard and cordless telephones (both pulse-dial and tone-dial), answering machines, and automatic dialers. The 1045 can be operated by a nontechnical person interested in finding out if a problem is caused by the telephone product or the telephone system. The 1045 checks the telephone-cords, ringing mechanism, dialing mechanism, mouthpiece, earpiece, and audio

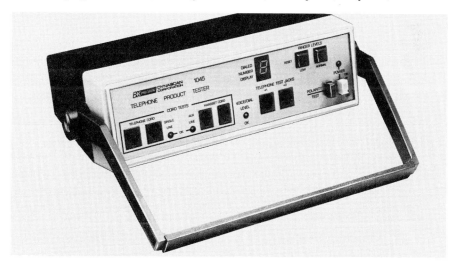

FIGURE 2-22. Model 1045 Telephone Product Tester (Courtesy of Dynascan Corporation)

amplifier. A second telephone is not necessary for test of telephones, but a second telephone jack is provided for test of answering machines.

Many telephone problems occur because of a faulty cord, so the manufacturer recommends that you always perform the cord test first. If the cords are found to be defective, the other tests should be conducted only after the cords are replaced. It is also recommended that all tests described here be performed on each telephone tested, so determine that all circuits of that telephone are operating normally.

The 1045 verifies performance of telephone products to be used in normally operating telephone systems. In some cases, due to variations in telephone systems, a telephone may check out good on the 1045, yet still fail to work with the telephone system. If this happens, a substitute telephone can be used to verify that the telephone system is operating. Or, a tester such as the B&K-Precision Model 1042 Telephone Line Analyzer (Sec. 2-6) can be used to check the telephone system.

2-5.1 Using the 1045 for Specific Tests

The remainder of this section is devoted to using the 1045 to perform specific tests on telephone products. The functions of the operating controls, as well as the operating procedures to perform specific tests, are described fully in the instruction manual for the 1045, and are not duplicated here. Instead, we concentrate on the *relationship* between the 1045 circuits, and the circuits within the telephone, during a particular test. This is done by showing both the 1045 test circuits (in block form), and the telephone circuits under test, for each test procedure.

A careful study of this relationship and the interconnections can help you to understand and troubleshoot the telephone circuits. If you know what is involved in testing a particular circuit, it often becomes quite clear what is wrong when that circuit does not perform properly. In any event, the test serves as a starting point for troubleshooting, as discussed in Sec. 2-7.

2-5.2 Operating Controls

Figure 2-22 shows the operating controls for the 1045. Reference to these controls is made throughout the remaining paragraphs of this section.

2-5.3 Cord Test

Figure 2-23 shows the test connections. This test is used to check a detachable handset (handset-to-desk unit) cord or the telephone (telephone-to-wall) cord. *If the cord is not detachable at both ends, the cord cannot be tested. Go on to other tests.*

If the telephone cord is found to be in good working order, any problem can then be isolated to the telephone or the telephone company equipment. If the cord

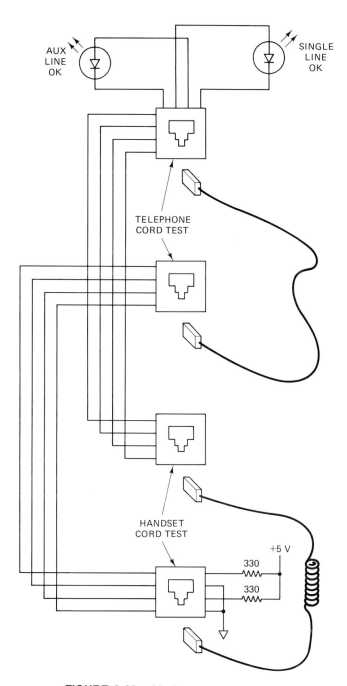

FIGURE 2-23. Model 1045 cord tests

is defective, other tests should still be carried out (after replacing the cord or cords) to ensure that there is no problem with the telephone itself.

Only one cord can be tested at a time. Do not plug both cords into the 1045 simultaneously. Handset and telephone cords are not interchangeable and should not be plugged into the wrong jacks.

For the cord test, both ends of a detachable telephone cord are plugged into the appropriate jacks, as shown in Fig. 2-23. Two sets of jacks are provided: the TELEPHONE CORD JACKS for checking detachable telephone cords (telephone-to-wall) and the HANDSET CORD JACKS for detachable handset cords. The two sets of jacks are wired in parallel, and either type of cord is tested in the same manner (but only one cord may be tested at a time).

A current is fed through the telephone cord. This current is taken from the +5-V supply through a 330-Ω resistor. At the other end of the cord, an LED is connected between one terminal of the cord jacks (one terminal and one LED for each pair of wires) and ground. One LED lights when a two-wire cord has continuity, with no shorts. Both LEDs light for a good four-wire cord. If either LED fails to light, the cord is defective, either shorted or no continuity.

1. Push the POWER switch ON. The POWER indicator lights when the 1045 is on.

2. To test the telephone cord for continuity or shorts, plug both ends of the cord into the TELEPHONE CORD TEST jacks. If the SINGLE LINE OK indicator lights, the cord is good for normal telephone operation. If the AUX OK indicator also lights, the cord is good for lighted telephone operation.

The SINGLE LINE OK indicator shows the condition of the two wires in the cord used for dialing, ringing, and conversation. The AUX LINE OK indictor shows the condition of the two wires in the cord that bring power to the night light or dial light in a telephone.

If the indicators fail to light, the cord is defective. Gently bend and squeeze the cord to check for intermittent continuity or shorts. If the indicators go out or flicker, the cord should be replaced.

3. Using the HANSET CORD TEST jacks, test the handset cord in the same manner as the telephone cord.

2-5.4 Dial Test

Figure 2-24 shows the test connections. This test applies to both tone-dial and pulse-dial telephones. If tones are heard at the earpiece when digits are pressed, the telephone is a tone-dial model. If a series of clicks are heard, it is a pulse-dial model. When numbers are dialed faster than two digits per second, the 1045 stores up to 16 digits in memory and releases the digits at two digits per second. If you wish to clear the DIALED NUMBER DISPLAY, press the RESET/LOW button. This clears the memory and places a "0" on the display.

For the dial test, the telephone to be tested is plugged into TELEPHONE TEST

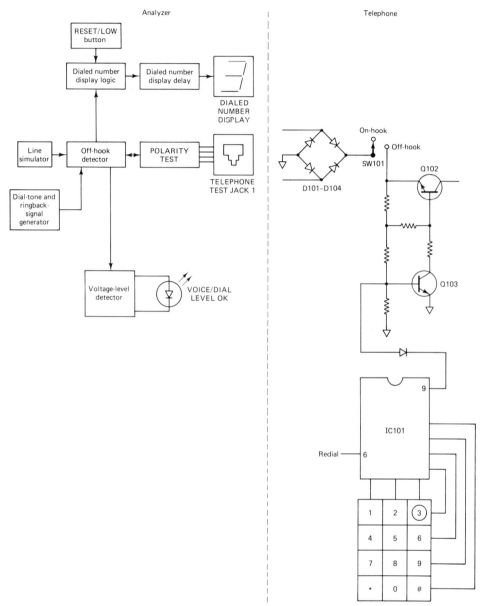

FIGURE 2-24. Model 1045 dial test

JACK 1 as shown in Fig. 2-24. The *off-hook detector* circuit connects the line simulator circuit and the *dial tone and ringback signal generator* to the telephone through the POLARITY TEST switch when the telephone is taken off-hook. The *line simulator* circuit applies -48 V through a 1.5 Ω resistance (to simulate a normal telephone line). The polarity test swtich reverses the d-c polarity to test the ability of the telephone to operate with either polarity. The dial tone and ringback signal generator circuit generates two tones (440 and 352 Hz), mixes them together to produce a precision dial tone, and feeds this tone to the telephone.

As soon as the first digit is dialed from the telephone, the dial tone is disconnected, and the tones or pulses are fed to the *dialed number display logic circuits* which decode the pulses or tones into a 4-bit binary logic signal. The 4-bit code is then fed to the *dialed number display delay*, which stores up to 16 dialed digits and then releases the digits at approximately two digits per second (so that the digits can be easily read from the DIALED NUMBER DISPLAY). The dialed number display delay feeds the code to a *decoder/driver*, which drives a seven-segment LED display.

1. Push the POWER switch ON. The POWER indicator lights when the 1045 is on.
2. Plug the telephone to be tested into the TELEPHONE TEST JACK 1. If a telephone is plugged into TELEPHONE TEST JACK 2, make sure that the telephone is on-hook throughout this test. Hang up both telephones (place the telephones on-hook).
3. Pick up the telephone under test (take the telephone off-hook) and listen for a dial tone. If no dial tone is present, this indicates that the telephone is not operating properly and should be repaired or replaced. (Trace the dial tone through the telephone audio circuits as described in Chapter 3.)
4. Press or dial each digit (be sure to use *all 10 digits*). With tone-dial telephones, it is also possible to test the "*" and "#" keys. The "*" key displays a decimal point (.) and the "#" key displays a bar (-) on the DIALED NUMBER DISPLAY.

Each digit should appear on the DIALED NUMBER DISPLAY in the same order as dialed. The decimal point lights momentarily each time a new digit is displayed so that if the same digit is dialed two or more times in sequence (for example, 933-3033), each individual digit can be distinguished. Also, each time a number is pressed on a tone-dial telephone, the VOICE/DIAL LEVEL OK indicator should light.

If the numbers do not appear on the DIALED NUMBER DISPLAY in the correct order, or if the VOICE/DIAL LEVEL OK indicator does not light each time a digit is pressed on a tone-dial telephone, the telephone dialing circuits are not oeprating properly. (Check the telephone dial circuits as described in Chapter 3.)
5. It is also possible to test the redial feature of a telephone. To do so, perform steps 1 through 4 of the dial test and then operate the redial feature for the telephone being tested. Each digit should again appear on the DIALED NUMBER DISPLAY in the same order as dialed. (If not, check the telephone redial circuits as described in Chapter 3.)

2-5.5 Ring Test

Figure 2-25 shows the test connections. This test applies to all types of telephones. *A word of caution before going on. Do not hold the telephone near your ear during the ring test. The ringing might be loud enough to cause hearing damage.*

For the ring test, the telephone to be tested is plugged into TELEPHONE TEST JACK 1. The *off-hook detector* circuit connects the telephone to the *line simulator, ring generator,* and *ring amplifier* through the POLARITY TEST switch when the telephone is on-hook. The *ring generator* produces a 20-Hz square-wave ring signal that is on for 1 second and off for 4 seconds. The square-wave ring signal is then passed through a low-pass filter which removes the harmonics and passes a 20-Hz sine wave.

The signal is then taken from one of two potentiometers, one set to give a normal ring voltage when the NORMAL RINGER LEVEL button is pressed, the other set to give a low ring voltage when the LOW RINGER LEVEL button is pressed. The ring signal (either normal or low) is then fed to the *ring amplifier*, which amplifies the ring signal to either 100 V (normal) or 45 V (low) and feeds the signal to the telephone.

When the telephone is taken off-hook, the off-hook detector circuit inhibits the ring generator. If a telephone is connected to TELEPHONE TEST JACK 2 and taken off-hook during the ring test, a ringback signal of 440 Hz, chopped by 20 Hz ringing at 1 second on and 4 seconds off, is produced by the *dial tone and ringback generator*.

1. Push the POWER switch ON. The POWER indicator lights when the tester is on.
2. Plug the telephone to be tested into TELEPHONE TEST JACK 1. Hand up the telephone (place the telephone on-hook).
3. Press the RESET/LOW RINGER LEVEL button. The telephone should ring until picked up (taken off-hook). If the telephone rings in this step, the telephone ringing circuits are good. If the telephone does not ring, go to step 4.
4. If the telephone does not ring in step 3, pick up the telephone (off-hook), and then replace the telephone (on-hook). Press the NORMAL RINGER LEVEL button. The telephone should now ring until picked up (off-hook). If the telephone still does not ring, there is a problem with the ring circuits. (Check the ring circuits as described in Chapter 3.)

If the telephone rings with normal ring voltage (100 V rms) but not with low ring voltage (45 V rms), this indicates that the telephone will

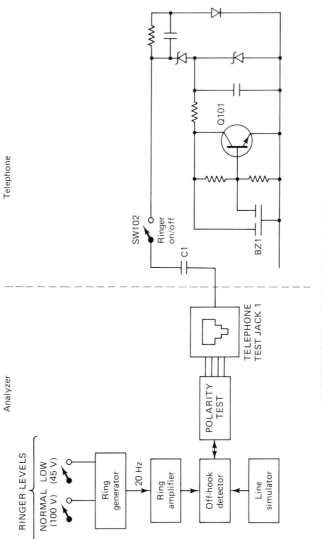

FIGURE 2-25. Model 1045 ring test

not work with a long-line situation (when the telephone is to be used many miles from a switching station).

If you wish to hear the ringback tone while the telephone is ringing, listen to a second telephone connected to PHONE TEST JACK 2.

Note that if a large number of telephones or telephone-type devices, are connected to a telephone line, this may prevent a good telephone from ringing. Refer to the note at the end of Sec. 2-3.8.

2-5.6 Voice-Level Test

Figure 2-26 shows the test connections. This test applies to all types of telephones. The telephone to be tested is plugged into TELEPHONE TEST JACK 1. The *off-hook detector* circuit connects the telephone to the *line simulator* and to the *voltage-level detector* through the POLARITY TEST switch when the telephone is taken off-hook. The voltage-level detector rectifies the tone-dial or voice signal

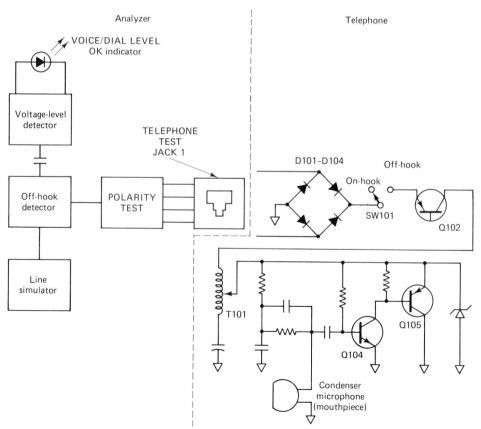

FIGURE 2-26. Model 1045 voice-level test

into a full-wave voltage which is fed to an op-amp. The op-amp feeds the voltage to the VOICE/DIAL LEVEL OK indicator LED, with a 5-V reference connected to the other end of the LED. If the voltage applied to the LED is at a level sufficiently above 5 V, the LED lights.

1. Push the POWER switch ON. The POWER indicator lights when the tester is on.
2. Plug the telephone to be tested into the TELEPHONE TEST JACK 1, and pick up the telephone (take the telephone off-hook).
3. Press the RESET/LOW RINGER LEVEL button to stop the dial tone. (It is important to turn off the dial tone while checking the VOICE/DIAL LEVEL OK indicator, since the dial tone causes the tester to read the voice level inaccurately.) The VOICE/DIAL LEVEL OK indicator should light, or flicker, when you talk into the telephone. If the indicators fail to light occasionally while you are talking, there can be a problem. (Check the audio circuits as described in Chapter 3.)
4. Press and hold the POLARITY TEST button while repeating steps 2 and 3 of the voice-level test. This reverses the polarity of the power applied to the telephone and should not affect operation of most telephones. Some telephones are not polarity guarded (telephones several years old). Check the schematic for polarity guarding, such as the full-wave diode bridge shown in Fig. 2-6.

2-5.7 Voice-Quality Test

Figure 2-27 shows the test connections. Two telephones of any type can be checked simultaneously. However, the test is most effective when one telephone is known to be in good working order (for both receiving and transmitting).

1. Push the POWER switch ON. The POWER indicator lights when the tester is on.
2. Plug the telephone to be tested into TELEPHONE TEST JACK 1. Connect the second telephone (preferably a known-good instrument) into TELEPHONE TEST JACK 2.
3. Pick up both telephones (take them off-hook). You should now be able to talk and listen as if a telephone call has been completed between two telephones.
    From telephone 1 you should be able to hear the person talking into telephone 2 (the known-good instrument). This checks the quality of the audio reception of telephone 1.
    From telephone 2 you should be able to hear the person talking into telephone 1. This checks the quality of the audio transmission of telephone 1.

B & K-Precision Model 1042 Telephone Line Analyzer   67

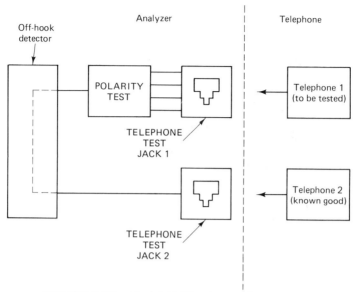

FIGURE 2-27. Model 1045 voice-quality test

4. Press and hold the POLARITY TEST button while repeating the test. This reverses the polarity of the power supplied to the telephone and should not affect operation.

2-5.8 Cordless Telephone Tests (Non-RF)

To test the non-RF functions of a cordless telephone, plug the base-unit a-c power cord into an a-c outlet, and plug the base-unit telephone cord into TELEPHONE TEST JACK 1. With these connections, the non-RF test procedures for cordless telephones are identical to the procedures for standard telephones, as described thus far in this section. However, be sure to follow the service literature for the telephone as well as the information in Chapter 4. The 1045 is not used to test the RF functions of cordless telephones.

2-6. B&K-PRECISION MODEL 1042 TELEPHONE LINE ANALYZER

Figure 2-28 shows the Model 1042 Telephone Line Analyzer. The 1042 quickly identifies a problem as external, or in the telephone itself, by testing the condition of the external telephone line, the telephone cord, the ring voltage level, the line voltage level, and the line polarity. The 1042 is simple to use, requires no power, and cannot disrupt the telephone line functions.

Figure 2-29 shows the complete schematic for the 1042, as well as typical test connections. As shown, the external line can be tested with, or without, a

68   The Basics of Telephone Equipment Troubleshooting and Repair

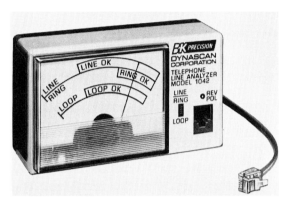

FIGURE 2-28. Model 1042 Telephone Line Analyzer (Courtesy of Dynascan Corporation)

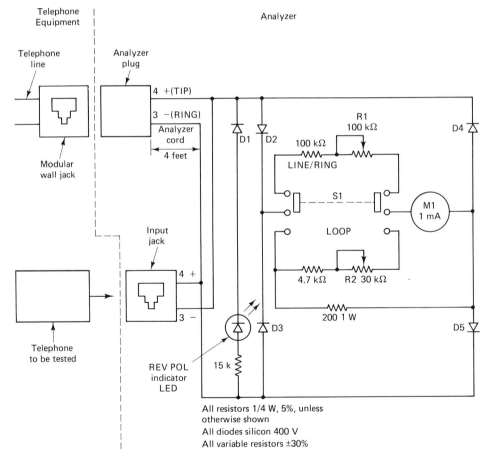

FIGURE 2-29. Model 1042 schematic and typical test connections

telephone. Therefore, if the line proves to be good by passing all the tests, but the telephone does not operate properly, you have pinpointed the problem to the telephone itself.

As shown in Fig. 2-29, the 1042 is essentially a meter, with provisions for connection to the external line and a telephone (in parallel with the line). The 1042 is so simple that you could make up a similar tester. However, you must calibrate the meter scale to reflect good and bad line conditions. By the time you have done this and have assembled all the parts, it will probably cost you more than the retail price of the 1042. However, we start this section by running through the calibration procedure. A careful study of this procedure will help you to understand how the 1042 tests a telephone line.

2-6.1 Analyzer Calibration

1. Set S1 to LINE/RING, and apply 40 V dc to the line input with the polarity shown in Fig. 2-30 (+ or plus to top or terminal 4, and - or minus to ring or terminal 3). Adjust R1 so that the meter needle just touches the LINE OK area on M1. With this setting, any line (battery) voltage 40 V or above produces a reading in the LINE OK area.

2. Reverse the polarity of the input voltage as shown in Fig. 2-31. The REV POL (reverse polarity) indicator LED should light, and M1 should read within 5% of step 1.

3. Apply a 40-V (rms) 20-Hz signal riding at a 48-V (dc) level to the line input as shown in Fig. 2-32. The reading should be within 5% of the lowest area on RING OK. With this setting, any 20-Hz ring voltage 40 V or above produces a reading in the RING OK area.

4. Set S1 to LOOP, and apply 20 mA dc to the line input as shown in Fig. 2-33. Adjust R2 so that the needle just touches the LOOP OK area on M1. With this setting, any loop current 20 mA or above produces a reading in the LOOP OK area.

2-6.2 Line Test

Figure 2-28 shows the operating controls and indicators.

1. Disconnect the telephone from the wall jack and plug the telephone into the input jack on the analyzer.

2. With the analyzer switch set to LINE/RING, connect the analyzer plug into the modular wall jack. The analyzer can now monitor line conditions, but with the telephone operating normally.

3. The meter should read in the LINE OK area. If not, first check that all telephone devices on the line under test are hung up (on-hook). If so, unplug one device at a time and watch the meter. If unplugging a

70   The Basics of Telephone Equipment Troubleshooting and Repair

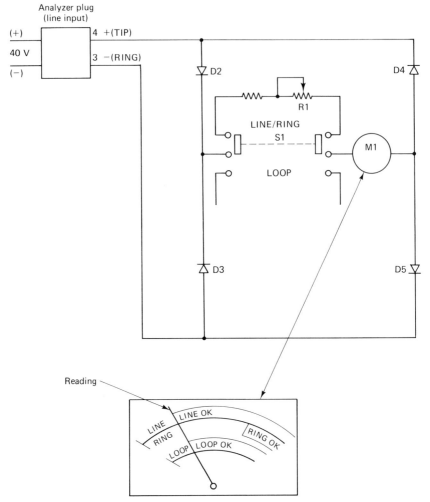

FIGURE 2-30.   Calibrating LINE/RING function

particular device causes the meter reading to increase significantly, it is likely that the device is defective. A properly operating device causes little line loading when on-hook. (If new wiring has been installed, particularly by a do-it-yourselfer, check the line carefully. Try disconnecting the new wiring to verify that the installation is not loading or shorting the incoming line.)

4. If the REV POL indicator lights, this indicates that the telephone line polarity is reversed from normal. If the telephone is polarity sensitive (mostly older telephones), reverse polarity can produce operating problems. However, all line tests can be made with either polarity (REV POL indicator on or off).

5. Once an acceptable line voltage reading is obtained, pick up each

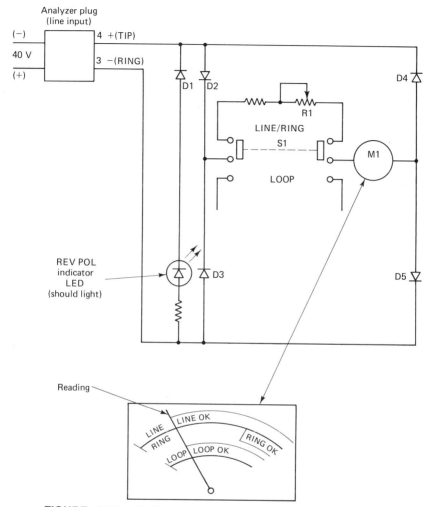

FIGURE 2-31. Calibrating LINE/RING function with reverse-polarity input

telephone device (go off-hook) and observe the meter. The reading should drop to near zero and return to the normal reading when the device is hung up (on-hook). If not, this indicates a possible problem with the telephone device.

2-6.3 Ring Test

Figure 2-28 shows the operating controls and indicators.

1. If the line voltage appears to be good (Sec. 2-6.2), dial a ringback number (if available) or have someone call you. (The telephone company will be

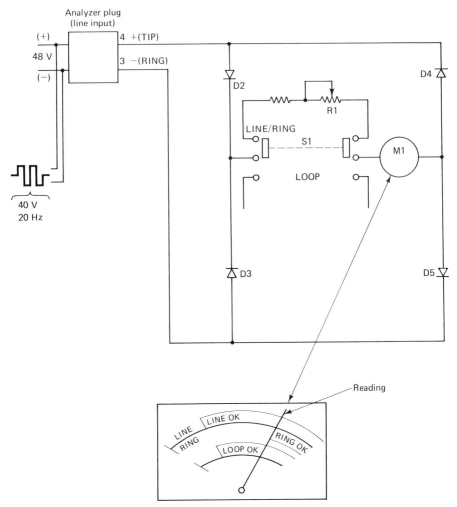

FIGURE 2-32. Calibrating RING OK functions

delighted to give you the ringback number; sure they will!) When the telephone rings, the meter reading should increase into the RING OK area for the duration of the ring, and the meter needle or pointer will vibrate during ringing.

Note that during the ring period, needle deflection should be at least 1/8 inch past the line-test reading (Sec. 2-6.2). This means that if (during the line test) the needle is in the RING OK area or very close, the needle should deflect an additonal 1/8 inch or more toward the high end of the scale.

2. If the telephone does not ring, but the meter reading increases into the RING OK area, the telephone ringer circuit may be defective. (Check the ring circuits as described in Chapter 3.)

3. If the telephone does not ring and the meter reading is not in the RING

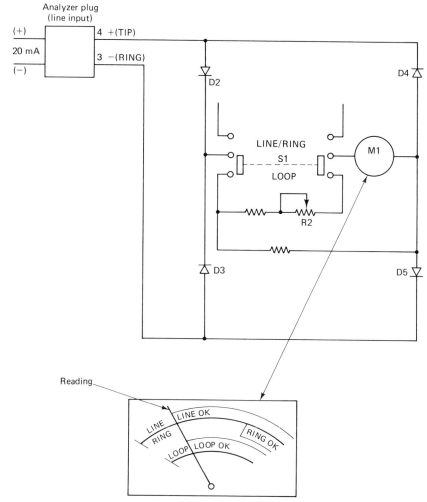

FIGURE 2-33. Calibrating LOOP OK function

OK area, it may be caused by a large number of telephone devices on the line. This can load the ringing signal to the point where the signal does not ring all of the telephone devices connected to one line. Add up the ringer equivalence numbers (sometimes called R.E.N.) indicated on each telephone device. If the total is five, the ring voltage may be on the borderline. (The telephone company usually guarantees to ring telephone devices with a maximum of five ringer equivalence numbers.)

If the ring voltage is not in the RING OK area, try checking the ring voltage at intervals. If total loading on an exchange is heavy, the ring voltage may be temporarily low. The telephone company master ring source has only a certain amount of power available.

4. Disconnect each telephone product, one at a time, and repeat the ring test. If removal of a particular telephone device clears a ring problem, that device is suspect.

5. If only one telephone is plugged into the line and a low ring reading is obtained, unplug the telephone from the analyzer jack and check if the ring voltage reading increases significantly. If there is a large increase in the reading, the telephone ringer circuit may be defective. However, if the increase is small and the meter reading remains in the marginal area, there is a ring voltage problem on the line. (Before you run screaming to the telephone company, make sure that any "modifications" performed by the user are not the cause of the problem.)

Note that when the meter reading is at the borderline of the RING OK area (indicated as the RING "?" area on some meter scales), the ring signal is very close to the minimum voltage guaranteed by the telephone company. Not all telephones will ring with this voltage.

2-6.4 Loop Test

Figure 2-28 shows the operating controls and indicators. This test verifies the conditon of the telephone line from the central office to the telephone jack in the home. This is done by showing that the line delivers a minimum current of 20 mA with the telephone devices connected.

If the line test (Sec. 2-6.2) and ring test (Sec. 2-6.3) produce low readings (after performing all additional checks suggested for these tests), the telephone line itself may be defective. Use the following procedure to pinpoint the problem.

1. With all telephone devices plugged into the line and the analyzer properly connected (analyzer plugged into the wall jack and a telephone plugged into the analyzer jack), set the analyzer switch to LOOP. If the meter reading is not in the LOOP OK area, check all telephone devices by unplugging each device, one at a time, and checking the meter reading.

2. Before calling the telephone company, repeat the loop test several times at 15-minute intervals to determine if the problem is temporary.

3. Do not leave the analyzer connected to the telephone line with the switch set to LOOP. This simulates an off-hook condition. Anyone calling the line number will get a busy signal.

2-6.5 Telephone Cord Test

Figure 2-28 shows the operating controls and indicators. With the telephone and analyzer connected as described for the line test (Sec. 2-6.2), the analyzer may indicate acceptable readings but the telephone may still not operate properly. If this is the case, the telephone cord (normally connected between the telephone and wall jack) may be defective. The analyzer can be used to test the cord as follows:

1. If the cord is detachable at both ends, make sure that both plugs are properly seated before going on with the test (this cures many telephone "problems"). If the cord is not detachable, the test cannot be used.

2. Transpose the connections described in Sec. 2-6.2. That is, connect the telephone cord between the wall jack and the input jack on the analyzer front panel; plug the analyzer cord into the telephone.

3. Set the analyzer switch to LINE/RING. Bunch the cord, squeeze, and release the cord, while checking for abrupt changes in the meter reading. Move the cord up and down, as well as back and forth, near each plug, while watching the meter. If there is no reading, or if the reading fluctuates during steps 3 and 4, the cord is defective and must be replaced. Of course, you could check the cord by replacement. However, this test using the analyzer is most convenient when a replacement cord is not readily available.

2-7. THE BASIC TROUBLESHOOTING APPROACH

Before we go on with troubleshooting techniques for specific telephone circuits, let us review the basic troubleshooting approach. It is assumed that you are already familiar with the basics of electronic troubleshooting, including solid-state troubleshooting. If not, and you plan to service telephones, you are in terrible trouble. Your attention is directed to the author's best-selling books: *Handbook of Practical Solid-State Troubleshooting* (Englewood Cliffs, N.J.: Prentice-Hall, Inc., 1971); *Handbook of Basic Electronic Troubleshooting* (Englewood Cliffs, N.J.: Prentice-Hall, Inc., 1977), and *Handbook of Advanced Troubleshooting* (Englewood Cliffs, N.J.: Prentice-Hall, Inc., 1983).

In the case of any telephone device, there are seven basic functions required for troubleshooting and repair.

First, you must study the telephone using service literature, user instructions, schematic diagrams, and so on, to find out how each circuit works when operating normally. In this way, you know (in detail) how a given telephone device should work. If you do not take the time to learn what is normal, you cannot tell what is abnormal. For example, many modern telephones have a redial feature (where the last number dialed is retained in memory and can be put onto the line by pushing a redial button). However, not all modern telephones have the redial feature, even though they contain a microprocessor IC with a built-in redial function. (Such is the case with the telephone circuits described in Chapter 1.) You can waste hours of precious time (money) trying to make a telephone redial when the redial circuit is not connected.

Second, you must know the function of, and how to operate, all of the telephone controls. It is assumed that you can operate a standard telephone, and possibly a cordless telephone. However, it never hurts to skim through the operating instructions of any telephone with which you are not completely familiar.

Third, you must know how to interpret service literature and how to use test equipment. Along with good test equipment that you know how to use, well-written service literature is your best friend. In general, telephone service literature is excellent as far as procedures (operation, adjustment, disassembly, reassembly) and illustrations (drawings and photos) are concerned. Unfortunately, telephone literature is often weak when it comes to descriptions of how circuits operate, why circuits are needed, and so on (the theory of operation). The "how it works" portion of much telephone literature is somewhat skimpy, or simply omitted, on the theory that telephone circuits are relatively simple and are contained in ICs (and therefore cannot be reached). That is why we concentrate so heavily on the "how it works" aspect of telephones, and the relationships between telephone circuits and test equipment during test/adjustment, in this book.

Fourth, you must be able to apply a systematic, logical procedure to locate troubles. Of course, a logical procedure for one type of telephone is quite illogical for another. That is why much of this book is devoted to an overall approach for troubleshooting any telephone device. It is assumed that you will study the service literature for any particular telephone equipment you are servicing. (If not, you will probably wish that you had taken the time to do so.)

Fifth, you must be able to analyze logically the information of an improperly operating telephone. The information to be analyzed may be in the form of performance, such as placing and receiving calls, redialing, and so on, or may be indications taken from test equipment, such as meter readings and waveforms monitored with an oscilloscope. Either way, it is your analysis of the information that makes for logical, efficient troubleshooting.

Sixth, you must be able to perform complete checkout procedures on a telephone that has been repaired. Such checkout may be only a simple operation, such as placing and receiving calls. At the other extreme, the checkout can involve complete readjustment of the telephone. In any event, some checkout is required after any troubleshooting.

One practical reason for the checkout is that there may be more than one trouble. For example, an aging part may cause high current to flow through a resistor, resulting in burnout of the resistor. Logical troubleshooting may lead you quickly to the burned-out resistor. Replacement of the resistor can restore operation. However, only a thorough checkout can reveal the original high-current condition that caused the burnout.

Another reason for after-service checkout is that the repair may have produced a condition that requires readjustment. A classic example of this is where components have been replaced in the transmitter or receiver circuits of cordless telephones. When any such components are replaced, complete check and possible realignment of the transmitter/receiver circuits is recommended. (We describe some typical procedures in Chapter 5.)

Seventh, you must be able to use proper tools and test equipment. The hand tools required for telephone service are essentially the same as for other types of electronic equipment (that involve digital logic circuits and ICs). The basic test equipment is also the same. However, there are many specialized items of test equipment for telephone service, such as those described in this chapter. It is recommended that you study this specialized test equipment carefully. It will help you to understand both the test equipment and the telephone circuits being tested.

In summary, before you tear into a telephone with soldering tool and hacksaw, ask yourself these questions: Have I studied all available service literature to find out how the telephone works (including any special features such as redial, call-waiting, etc.)? Can I operate the telephone controls properly? Do I really understand the service literature and can I use all required test equipment (especially the specialized test equipment for telephones)? Using the service literature and/or previous experience on similar telephones, can I plan out a logical troubleshooting procedure? Can I analyze logically all the results of operating checks as well as checkout procedures involving test equipment? Using the service literature and/or experience, can I perform complete checkout procedures on the telephone, including realignment, adjustment, and so on, if necessary? Once I have found the trouble, can I use common hand tools to make the repairs? If the answer is no to any of these questions, you simply are not ready to start troubleshooting any telephone, modern or otherwise. You had better start studying!

3

THE BASICS OF TROUBLESHOOTING CORDED TELEPHONES

This chapter is devoted to the basics of troubleshooting corded telephones. As discussed throughout this book, the first step in troubleshooting is to test the instrument against known standards. In Chapter 2 we describe a complete set of tests for typical corded telephone circuits using specialized test equipment. In this chapter we describe how to perform similar tests without specialized test equipment. We also make reference to corresponding tests in Chapter 2. By comparing the procedures of the two chapters, you will quickly realize the advantages of specialized test equipment, particularly if you plan on servicing telephone equipment regularly.

3-1. TESTING THE TELEPHONE LINE

If you are servicing a telephone in the shop, where a known-good telephone line is available, you need not check the line before going on with other tests. In the field it is often a good idea to check the line first. This can be done by checking the line with a known-good telephone, or with a line tester, as described in Sec. 2-6. Either way, if the line proves to be good, you then know that there is a problem in the suspect telephone. Of course, if the line is bad, there is no point in going on until the line is cleared (either in the house wiring or the external telephone line). Note that if the telephone "fails" immediately after do-it-yourself wiring, you have an excellent starting point for troubleshooting.

3-2. TESTING CORDS

Always check the cords first, long before you launch into an attach on the telephone circuits. The cords take the worst beating of any telephone equipment, except for those few cases where the telephone or handset are bounced off walls and floors. So it is logical that the cords fail first. Look for signs of excessive wear, cuts, knicks, frayed wires, and so on.

The simplest way to check the cords, either the *handset cord* or the *telephone cord* shown in Fig. 3-1, is by substitution. If you plan on servicing telephone equipment on a grand scale, you should have at least one set of known-good cords (handset and telephone) for substitution. If the known-good cords solve the problems, you have pinned down the source of trouble.

If you do not have substitute cords, you can make a *continuity check* of the suspect cords as shown in Fig. 3-2. Check each conductor in the cords for continuity by connecting an ohmmeter at both ends of the cord. The resistance should be zero, or very near zero. If one or more of the conductors shows a high resistance (or a resistance substantially different from the other conductors), there is an open or partially open condition.

To perform the continuity check, it is necessary to disconnect the cord from *both ends*. If the cord is permanently attached at one end (typical for older telephones) it may be necessary to open the phone and disconnect the cord wiring (if the wiring is not soldered in place).

Next check the resistance from each conductor to all other conductors. If there is a low resistance between conductors, you have a short or partial short.

Be sure to bend and twist the cords when checking continuity and for shorts. This will show up an intermittent condition, but also brings up a problem.

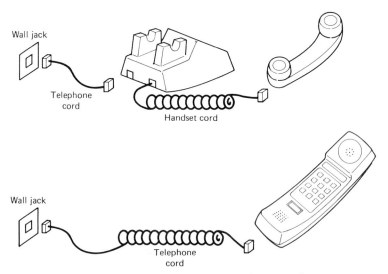

FIGURE 3-1. Typical handset and telephone cords

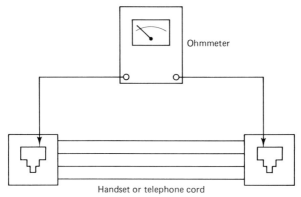

FIGURE 3-2. Checking handset and telephone cord continuity with an ohmmeter

At best, it is quite clumsy to bend and twist a cord while holding ohmmeter prods at each end, especially when the conductor terminals are quite small in a modular telephone plug. That is why the testers and analyzers described in Chapter 2 include provisions for checking cords.

You can make up a similar test circuit, as shown in Fig. 3-3. You need a power source, resistors, LEDs, and modular jacks or receptacles. Keep in mind that such a test configuration proves only that you have continuity but does not establish the possibility of shorts between conductors, as described in Secs. 2-3.6 and 2-5.3. So substitution is the quickest and easiest, particularly when you consider that you must replace the cord if the test proves the cord defective. Do not waste time trying to repair a cord unless there is a desperate emergency.

3-3. TESTING THE RINGER CIRCUITS

Once you have established that the line and cords are good, the next logical step is to check the ringer circuits. Of course, if there is no complaint about the ringing function of the telephone, the ring circuits need not be checked immedi-

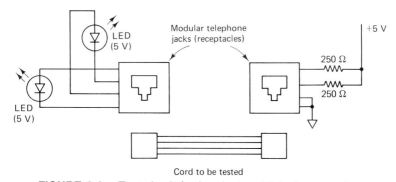

FIGURE 3-3. Test circuit for handset and telephone cords

82 The Basics of Troubleshooting Corded Telephones

ately after the cords. However, the ring circuits must be checked at some point. A failure in the ring circuits (such as leaking or shorted coupling capacitors) can cause an apparent failure to other circuits, as we discuss in the following paragraphs. So let us check the ring circuits next.

3-3.1 Basic Ring-Circuit Test

The obvious test of the ring circuits is to apply a ring voltage of the correct amplitude (40 V rms minimum) and frequency (20 Hz) as shown in Fig. 3-4, and see if the telephone rings. Unfortunately, it is not that simple.

First, the ring voltage should be variable to provide a proper test of the ring-circuit *threshold voltage*. While the telephone company generally supplies 80 to 130 V rms to the line, other devices on the line (and the line resistance) can cause the ring voltage to drop as low as 40 V rms. If the telephone does not ring with a ring voltage of 40 V rms, the ringer circuit is probably defective. (You may have a borderline case where the telephone rings with something between 40 and 45 V rms applied. However, if the ring circuits need 45 V rms or higher, you have a problem).

Next, and more important, the ring voltage must be applied only when the telephone is on-hook. If you apply the ring voltage to an off-hook telephone, you will probably blow the remaining circuits (in addition to blowing your eardrums). Unfortunately, even if the handset is in place on the telephone, and the instrument appears to be in the on-hook condition, the circuits may be defective (on-hook/off-hook switch jammed to the off-hook position). This is why the analyzers discussed in Chapter 2 have an off-hook detector circuit to cut off the ring voltage when an off-hook condition is detected. The telephone exchanges have similar off-hook detector systems. These circuits are not easily duplicated, which is another reason for using the specialized test equipment.

With the Chapter 2 analyzers, the ringing stops immediately when the telephone is lifted off-hook (as is the case when the telephone is connected to a

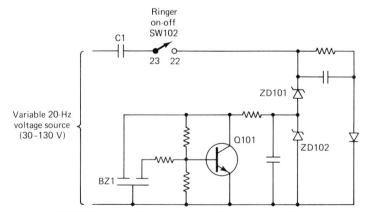

FIGURE 3-4. Basic ring-circuit test connections

telephone line). If the telephone continues to ring, the telephone is not coming off-hook properly, and should be checked. Before you go inside, make sure that nothing prevents the hook switch from moving. (Also look for obvious problems, such as the ringer on–off switch SW102 in Fig. 3-4 being set to off, or open.)

If the telephone does not ring, with a proper ring voltage applied as shown in Fig. 3-4, the next step is to trace the ring voltage through the circuits, on a point-to-point basis. A good place to start is on both sides of SW102, at terminals 22 and 23. Then go on to the ringer (buzzer) BZ1 and transistor Q101. Pay particular attention to zeners ZD101 and ZD102.

If the telephone does not ring when being tested with an analyzer such as described in Secs. 2-3.8 and 2-5.5, check the VOICE/SIGNAL LEVEL indicator on the analyzer. The indicator should flicker when the ring button is pressed. If not, the analyzer may have cut off the ring voltage because an off-hook condition has been sensed by the analyzer off-hook detector. For example, if C1 in Fig. 3-4 is leaking or shorted, the on-hook impedance of the telephone may be low (say around 250 Ω), causing the detector to cut off the ring signal. (This same condition can cause the telephone company circuits to cut off the ring signals.)

Before you decide that there is an impedance problem in the ringer circuits, check the wiring. Some older telephones are wired for three-wire operation, where two wires are used for voice and dialing and a third wire is used for ringing. Usually, this condition can be corrected by connecting the telephone ring wire (yellow) to one of the other two wires (typically, the green wire). This problem does not exist on modular telephone systems. In any event, do not attempt to rewire the telephone line to make a particular telephone operate properly. The telephone company takes a dim view of such actions.

3-3.2 Tracing Ring-Circuit Signals

Now let us trace voltages or signals through some typical ring circuits, such as those shown in Fig. 3-5. Note that the ring circuits of Fig. 3-5 are somewhat more complex than those shown in Fig. 3-4, even though overall operation of both circuits is essentially the same. So we will go through the circuit functions before we describe the signal tracing procedure.

In the circuits of Fig. 3-5, the 20-Hz ring voltage is applied to the ring circuits through C701–C702 and D701–D704. Capacitive coupling is used to prevent the telephone line d-c voltage (battery power) from turning the circuits on. The rectified d-c output from D701–D704 is used to power the ring circuits. Zener Z701 and transient suppressor C722 prevent small signals, such as dial tones or audio, from triggering the circuits.

Note that ringer or buzzer BZ701 can be disabled by opening ringer on/off switch S801. Also, when the "talk" mode is selected, a voltage is applied through R736 to disable the ring circuits (by driving TR710 and TR709 into saturation).

The ringing signal (of at least 40 V rms at 20 Hz) produces an output from D701–D704 that exceeds the Z701 zener voltage turn-on point and provides

84   The Basics of Troubleshooting Corded Telephones

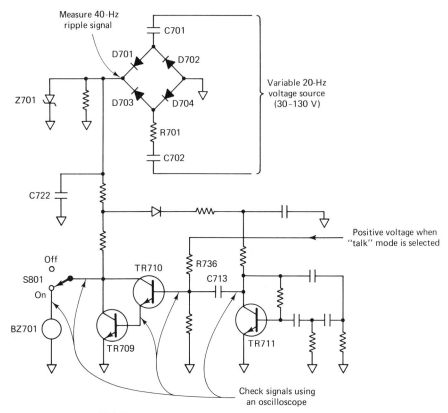

FIGURE 3-5. Tracing ring-circuit signals

power to oscillator TR711 and amplifiers TR709/TR710. Amplified oscillations are applied to ringer BZ701 through S801.

The first step in tracing the ring-circuit signals is to apply a 20-Hz signal at about 100 V rms to the telephone line input as shown in Fig. 3-5. Make absolutely certain that the telephone is disconnected from the telephone line and is on-hook. As we have discussed, if the telephone is not on-hook, you could destroy the remaining telephone circuits. Similarly, if the telephone is not disconnected from the line, your ring voltage may prove annoying to the technicians at the telephone exchange!

Measure from the junction of D701 and D703 to ground with an oscilloscope. (You are measuring across zener Z701 and resistor R722.) Because D701–D704 form a full-wave bridge rectifier, the 20-Hz ring signal is converted to a 40-Hz ripple signal.

If the signal is absent, check for an open C701, C702, R701 or shorted C722 or Z701. Note that if C701, C702, or C722 are shorted or leaking, this can change the line impedance and cause the analyzer to cut off the ring voltage.

If the signal is normal across Z701, check for signals across ringer or buzzer BZ701 using an oscilloscope. Since the 40-Hz ripple acts as a power source for oscillator TR711, the signal should be interrupted at a 40-Hz rate.

If you get a signal across BZ701 but no ring, BZ701 is suspect. If there is no signal across BZ701, check the contacts of switch S801, as a first step . If S801 is good, check for a signal from oscillator TR711 at both sides of C713. Then check amplifiers TR709/TR710 and the associated circuit parts.

3-4. TESTING THE DIALING CIRCUITS

With the telephone line, cords, and ringer circuits checked out, the next order of business is to check the dialing circuits. Again, if there is no complaint about the dialing circuit, you may want to go directly to the circuit involved (such as the audio circuits) in the problem. However, in most electronic telephones, the audio signals must pass through transistors in the dialing circuit. As a result, any problem in the dialing circuits can cause apparent failure in other circuits, particularly the audio circuits.

The dialing circuits of both rotary-dial and pulse-dial telephones can be checked with an oscilloscope. Tone-dial telephones require some way to measure the frequencies of the tone pairs generated as each digit is dialed. This can be done with an oscilloscope and signal generator, but a frequency counter may prove easier. (Keep in mind that you must measure a *pair of tones*, and both tones must be on-frequency.) An analyzer can be used to test all three types of dialing circuits. The analyzer decodes both tone and pulse signals, and displays the actual number dialed on a DIALED NUMBER display.

No matter what method is used, *each digit should* be tested for the correct number of pulse or tone frequencies. In the case of tone-dial telephones, the level of the dialing signal must be checked.

3-4.1 Rotary-Dial Circuit Tests

Rotary-dial telephones use a set of contacts that are mechanically opened and closed to dial numbers. When the dial is released, contacts open and close accordingly. For example, for the digit "3" the contacts open and close three times, interrupting the telephone line three times. A mechanical governor controls the speed of rotation, and thus the period and repetition rate of the dial pulses.

The dial pulses can be checked by monitoring the telephone line with an oscilloscope at some convenient point in the circuit. If an analyzer is used, the oscilloscope can be connected at the LINE SCOPE jack, which is connected across the telephone line. Keep in mind that the amplitude of the pulses depends on the line voltage at the point of measurement.

Usually, if a rotary-dial telephone does not dial, it is because the contacts are bent, oxidized, or corroded. Oxidized contacts may be cleaned (carefully). However, if the contacts are bent or corroded, they should be replaced. In many

86 The Basics of Troubleshooting Corded Telephones

rotary-dial telephones, the entire mechanical dial assembly can be replaced as a unit.

3-4.2 Pulse-Dial Circuit Tests and Signal Tracing

Figures 3-6 and 3-7 show the dialing circuits of typical pulse-dial telephones connected for test using an oscilloscope. Sections 2-3.7 and 2-5.4 describe similar test procedures using analyzers.

If a pulse-dial telephone fails to dial, the first place to check is at the dialing IC (at pin 9 of IC101 in the circuit of Fig. 3-6 and at pin 18 of IC701 in Fig. 3-7). The IC output pulses a corresponding number of times for each digit pressed (one pulse for the "1" digit, eight pulses for the "8" digit, etc.).

If there are no dialing pulses from the IC, look for problems in the power and ground connections. Also look for any special connections that are required to place the dialing IC in operation.

As an example, IC101 in Fig. 3-6 is turned on only when pin 12 is returned to ground through hook switch SW101 (in the off-hook condition). This prevents dialing pulses from being generated by IC101 when the telephone is on-hook (say, if any of the keys or pushbuttons are pressed accidentally in the on-hook condi-

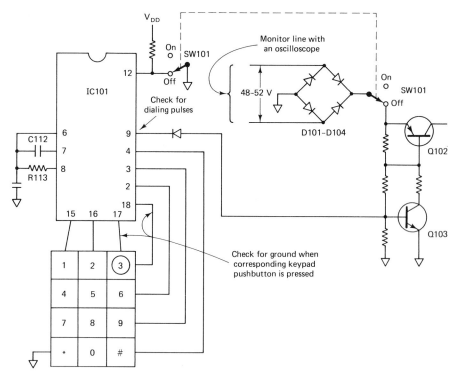

FIGURE 3-6. Pulse-dial-circuit tests and signal tracing

Testing the Dialing Circuits   87

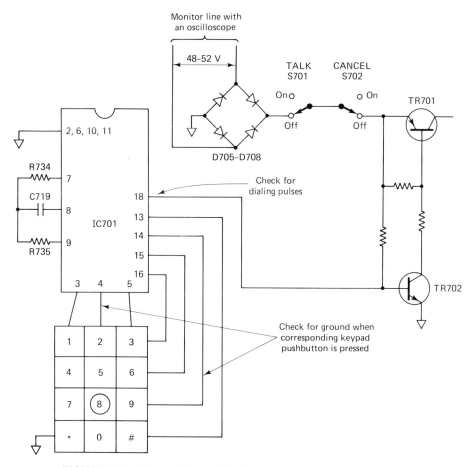

FIGURE 3-7. Pulse-dial-circuit tests and signal tracing (alternate circuit)

tion). If pin 12 of IC101 is not returned to ground (SW101 contact defective, bad wiring between IC101-12 and ground, etc.), IC101 does not produce any dialing pulses.

If there are dialing pulses from the IC but they are not clocked out at a steady rate, look for problems in the timing circuits (C112 and R113 in Fig. 3-6, C719, R734, R735 in Fig. 3-7). These components control the storage circuits in the dialing IC so that the dialing pulses are produced at a steady rate, no matter how quickly or slowly they are generated at the keypad (pushbuttons).

If there are dialing pulses from the IC, but not for all digits, check the corresponding wiring between the keypad and dialing IC. For example, if the digit "3" does not produce three dialing pulses from pin 9 of IC101 in Fig. 3-6, check the wiring from pins 17 and 18 of IC101 to the keypad (row 1 and column 3).

88   The Basics of Troubleshooting Corded Telephones

Typically, when a keypad pushbutton is pressed, the corresponding row and column wires are returned to ground. You can monitor this condition with an oscilloscope (or a logic probe). If the keypad and wiring appear to be good but the dialing pulses are not correct, the next step is to replace the dialing IC. However, it is generally much easier to check the keypad and wiring first if the dialing pulses from the IC appear abnormal.

If the dialing IC produces the correct number of pulses but the telephone fails to dial, the problem is most likely in the switching transistors (Q102/Q103 in Fig. 3-6, TR701/TR702 in Fig. 3-7). These transistors take the place of the mechanical contacts found on rotary-dial telephones, "opening" and "closing" the line in response to pulses from the dialing IC (for the same period and at the same repetition rate as the pulses).

When monitoring the line with an oscilloscope, check that the switching transistors "open" and "close" the line completely for each pulse. For example, using an oscilloscope with d-c coupling, note the d-c reference level with the telephone on-hook as shown in Fig. 3-8. Then note if the line switches all the way to the reference level for each dialing pulse.

Figure 3-8 shows the effects of leakage in the switching transistors. Such leakage can prevent the transistor from completely turning off during the dial pulse. Typically, leakage of as little as 1 mA can prevent some or all of the dial pulses from being recognized by the telephone exchange (or the analyzer if used).

Leakage in the switching transistors is usually a greater problem when the telephone line voltage is high. That is, the switching transistors may dial properly when the telephone line is 48 V or less, but may not completely shut off with higher voltages. Note that the analyzer described in Sec. 2-3 has a nominal 52-V

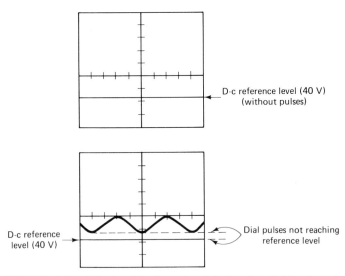

FIGURE 3-8. Effects of leakage in dial-circuit switching transistors

telephone line voltage rather than 48 V. This increase in voltage makes it easier to reject telephones with marginal switching transistors and to find those with impending failure.

Telephones with leakage problems in the switching transistors often appear to break down gradually, with the problem first occurring when the telephone line voltage is high. So, when using the test circuits of Figs. 3-6 and 3-7, check the dialing circuits with a telephone line voltage of 48 V. Then increase the line voltage to 52 V (if practical), while monitoring the line for any evidence of leakage (as shown in Fig. 3-8).

3-4.3 Tone-Dial Circuit Tests and Signal Tracing

Figure 3-9 shows the dialing circuits of a typical tone-dial telephone ("Touch Tone" or DTMF) connected for test using a frequency counter. Sections 2-3.7 and 2-5.4 describe similar test procedures using analyzers. In the circuit of Fig. 3-9, the tone-pair outputs of the dialing IC (IC404) are applied to the telephone line (and to the telephone speaker) through audio circuits.

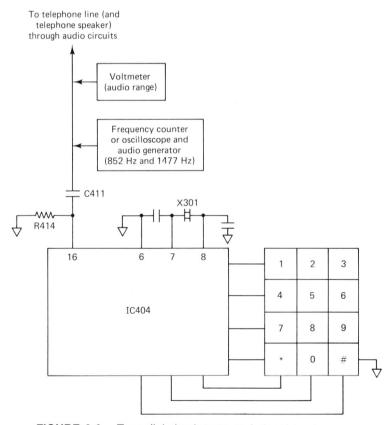

FIGURE 3-9. Tone-dial-circuit tests and signal tracing

The frequency of pulses from dialing IC404 are controlled by the 3.58-MHz crystal X301. (The precise frequency of X301 is 3.759545 MHz, which happens to be the color-TV burst frequency.) The frequencies produced by IC404 for each digit of the keypad are shown in Fig. 1-2.

If the tone-dial telephone fails to dial, check for tone-pair outputs at pin 16 or IC404. There should be a corresponding pair of tones at IC404-16 each time a keypad pushbutton is pressed. For example, if the "9" digit is pressed, the tone-pair output should be 852 Hz and 1477 Hz.

In addition to checking for the correct tone-pair frequencies, the signal level or amplitude should be measured. Of course, the actual signal level depends on the particular circuits. (Check the service literature!) However, as a guideline, each tone should be at an amplitude of at least 0.05 V rms on the line.

If you are using an analyzer, the VOICE/SIGNAL LEVEL indicator LED lights if the tone level is good, making it unnecessary to measure the actual amplitude of the tone-pair signals. Similarly, with an analyzer, the DIALED NUMBER display decodes the tone pairs and shows the actual number dialed. This makes it unnecessary to measure the tone-pair frequencies produced by the dialing IC.

3-5. TESTING THE AUDIO CIRCUITS

Generally, the audio circuits should be checked last, after the telephone line, cords, ringer and dialing circuits are proven good. This is because problems in other circuits can make the audio circuits appear defective. For example, defective cords can interrupt the audio, and a bad telephone line can attenuate the audio, as can leakage in the dialing circuit switching transistors or the ring-circuit coupling capacitors.

The audio circuits of electronic telephones can be checked in the same way as any audio circuit. Either *signal tracing* or *signal injection* can be used. However, signal injection with an audio generator is usually the simplest.

No matter what method is used, the level and quality of audio should be checked *in both directions* (to and from the telephone line). The voice at the earpiece should be loud enough so that it is easily understood. Outgoing voice should also be loud enough and clear enough to be easily understood at the earpiece of the other telephone. Also, the sidetone volume should be at a level where the person speaking into the telephone does not feel that they must shout or whisper.

3-5.1 Audio-Circuit Test and Signal Tracing

Figures 3-10 through 3-13 show the audio circuits of typical electronic telephones connected for test. Sections 2-3.9, 2-3.10, 2-5.6 and 2-5.7 describe similar test procedures using analyzers.

Note that the incoming signals to the telephone circuits of Figs. 3-10 and 3-

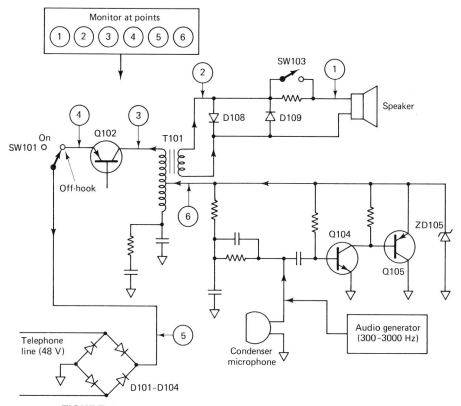

FIGURE 3-10. Audio-circuit outgoing-signal test connections

11 do not have amplification, as do incoming signals to the circuit of Figs. 3-12 and 3-13. Let us start with the Figs. 3-10 and 3-11 circuits first.

Although the actual audio signal level from a telephone depends on the particular circuits, most electronic telephones produce an output to the line of at least 0.1 V rms with normal voice levels. If not, transistors Q104/Q105 and the associated circuit parts are suspect.

Start by injecting an audio signal (between 300 and 3000 Hz) at the microphone as shown in Fig. 3-10. The signal should be heard on the speaker and should be available at hybrid transformer T101, Q102, SW101, and D101–D104. If the audio signal is absent or abnormal (very low, distorted, etc.) at any of these points, you have located the problem. Keep in mind that Q102 must be turned on and SW101 must be off-hook for audio (incoming or outgoing) to pass. Check the voltages at the transistor elements against the values shown in the service literature. Then check the circuit values.

Next, inject the same audio signal at the speaker and work back to diodes D101–D104, as shown in Fig. 3-11. Again, the signal should continue to be heard on the speaker as the audio generator is moved from point to point toward

92   The Basics of Troubleshooting Corded Telephones

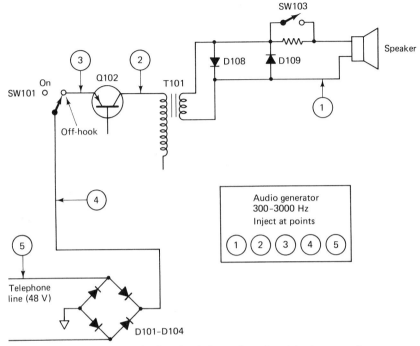

FIGURE 3-11. Audio-circuit incoming-signal test connections

the diodes. Again, Q102 must be turned on and SW101 must be in the off-hook position for audio to pass.

In the circuit of Fig. 3-12, inject the audio signal at the microphone as shown. Check that the signal is heard on the speaker and is available at TR704–TR708, T701, TR701, S701, S702, and D705–D708.

In the circuit of Fig. 3-13, inject the audio signal at the speaker and work back to diodes D705–D708 as shown. The signal should continue to be heard on the speaker as the audio generator is moved toward the diodes.

In both circuits (Figs. 3-12 and 3-13), switches S701 and S702 must be closed and TR701 must be turned on for audio to pass to the line diodes D705–D708.

Testing the Audio Circuits 93

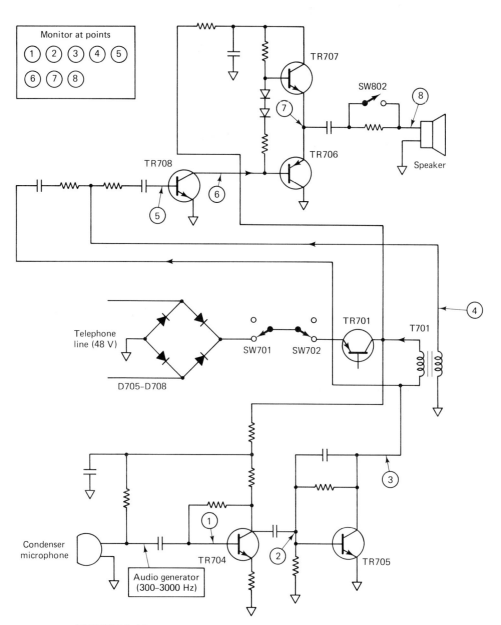

FIGURE 3-12. Audio-circuit outgoing-signal test connections (alternate circuit)

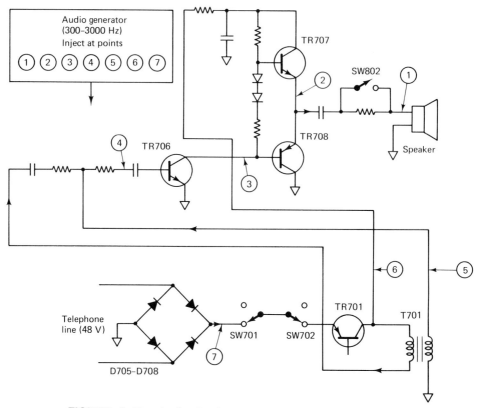

FIGURE 3-13. Audio-circuit incoming-signal test connections (alternate circuit)

4

THE BASICS OF TROUBLESHOOTING CORDLESS TELEPHONES

This chapter is devoted to the basics of troubleshooting cordless telephones. Examples of detailed, circuit-by-circuit, troubleshooting for cordless telephones are given in Chapter 5. As discussed throughout this book, the first step in troubleshooting is to test the instrument against known standards. In Chapter 2 we describe a complete set of tests for typical cordless telephone circuits using specialized test equipment. In Chapter 3 we describe tests for the non-RF portions of a cordless telephone using both an analyzer and standard test equipment. In this chapter we describe those tests that are required as a preliminary step for troubleshooting and then leave the choice of test equipment up to you. By comparing the procedures of the three chapters, you will quickly realize the advantages of specialized test equipment, particularly if you plan on servicing telephone equipment regularly.

Most of the procedures described in this chapter are supplemented with block diagrams, rather than schematics, to simplify the descriptions. Detailed circuit descriptions, as well as step-by-step circuit troubleshooting procedures, for typical cordless telephones are described in Chapter 5. The Chapter 5 descriptions are supplemented with schematic diagrams of typical cordless telephone circuits.

4-1. TESTING THE TELEPHONE LINE AND BASE-UNIT CORDS

As discussed in Chapter 3, both the telephone line and cords (for the base unit) should be tested before extensive troubleshooting of any telephone. The procedures described in Sec. 3-1 and 3-2 also apply to cordless telephones and should be performed first.

4-2. PRELIMINARY CHECKOUT

Once the telephone line and base-unit cords have been checked, the cordless telephone should be subjected to a preliminary checkout before you go inside for detailed troubleshooting as described in Chapter 5. This checkout, which takes only a few minutes, confirms any trouble symptoms reported by the customer and gives you a starting point for isolating the trouble to a specific telephone circuit or function.

If the telephone fails one or more of the tests, you have isolated the trouble to probable circuits or sections. If the telephone passes all the tests, you know that the cordless system is good, at least for short-range operation. You can then go on to some additional tests which verify good performance over longer ranges, under conditions of severe interference, and so on.

The following paragraphs describe the four basic tests for preliminary checkout of any cordless telephone, including those covered in Chapter 5.

4-2.1 Ring Test

Apply power to the base unit and connect the base unit to a telephone jack. Leave the portable unit in standby (on-hook). Apply a ring voltage or signal to the telephone line and see if the portable unit rings.

Sections 2-3.8 and 2-5.5 describe test connections using an analyzer, while Section 3-5 describes tests using conventional test equipment.

Start with a 100-V 20-Hz ring signal. If the telephone does not ring, go to the ring test troubleshooting of Sec. 4-3. If the telephone rings with 100 V, reduce the ring signal to 40 V and repeat the test. The telephone should ring with 45 V, and must ring with 40 V. If not, refer to Sec. 4-3.

4-2.2 Dial-Tone Test

Remove the 20-Hz ring signal *before* starting the dial-tone test. Set the portable unit to TALK and listen for a dial tone. If there is no dial tone, refer to the troubleshooting described in Sec. 4-4. If you get a satisfactory dial tone, go on with the dial test.

4-2.3 Dial Test

Dial each digit on the keypad, and check that the portable unit produces the correct pulses or tone pairs. Sections 2-3.7 and 2-5.4 describe test connections using an analyzer, while Section 3-4 describes tests using conventional test equipment.

If one or more of the digits does not produce the correct pulses or tone pairs, refer to the troubleshooting described in Sec. 4-5. If you get satisfactory dialing functions, go on with the voice test.

4-2.4 Voice Test

Speak into the portable-unit mouthpiece and check for satisfactory sidetone level and overall voice quality. Sections 2-3.9, 2-3.10, 2-5.6, and 2-5.7 describe test connections using an analyzer, while Sec. 3-5 describes tests using conventional test equipment.

If there appears to be a problem in the audio circuits, refer to the troubleshooting described in Sec. 4-6. If voice is satisfactory, the cordless telephone is operational, at least at short ranges. However, there may be a problem at longer ranges (refer to Sec. 4-7).

4-3. RING-CIRCUIT BASIC TROUBLESHOOTING

Figure 4-1 is the basic troubleshooting tree for ring-circuit problems in cordless telephones. Figures 4-2 and 4-3 show the circuits involved for cordless telephones operating in the 1.7/49-MHz and 46/49-MHz ranges, respectively.

As shown in Fig. 4-1, the first step is to apply a 20-Hz ring signal or voltage to the telephone line. As shown in Figs. 4-2 and 4-3, the 20-Hz ring detector recognizes the ring signal, and turns on the base-unit RF transmitter. The generator modulates the RF carrier with a ring signal (typically, a specific audio frequency in the range 700 to 1500 Hz, but check the service literature).

The portable-unit receiver detects the RF signal. A narrow band filter

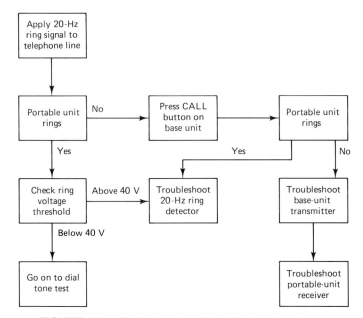

FIGURE 4-1. Basic ring-circuit troubleshooting tree

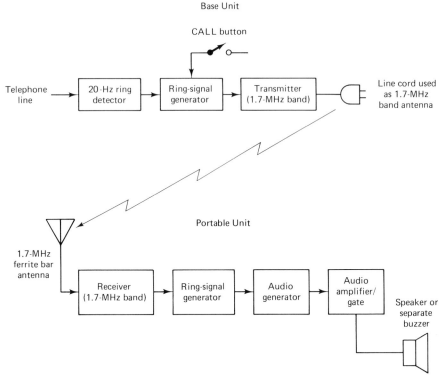

FIGURE 4-2. Basic ring circuits for 1.7/49-MHz cordless telephones

(within the ring signal detector circuit) detects the correct ring frequency and feeds an audio tone to the speaker. Most cordless telephones use the same speaker as that used in voice communications. However, some cordless telephones use a separate buzzer or ringer.

If the telephone rings, all of the circuits shown in Figs. 4-2 and 4-3 are operating properly. If not, one of the circuits is defective. Go on to Chapter 5.

If the ring test is satisfactory at 100 V, gradually reduce the ring-signal voltage. Note the approximate voltage at which ringing stops. Some cordless telephones may ring with a ring signal of only a few volts. However, most cordless telephones stop ringing when the ring signal is below 40 V. In any event, if the ring threshold is higher than 40 V, check the 20-Hz ring detector circuit in the base unit (not the ring-signal detector in the portable unit).

Most base units have a CALL button which can be pressed to ring the portable unit without a 20-Hz ring signal applied to the telephone line. As shown in Fig. 4-1, if the portable unit does not ring with a 20-Hz ring signal applied to the base unit, press the CALL button and check if the portable unit rings.

In most cordless telephones, the CALL button activates the base-unit RF transmitter and applies ring-signal modulation, causing the portable unit to ring. In this case, since there is no 20-Hz ring signal applied to the telephone line and

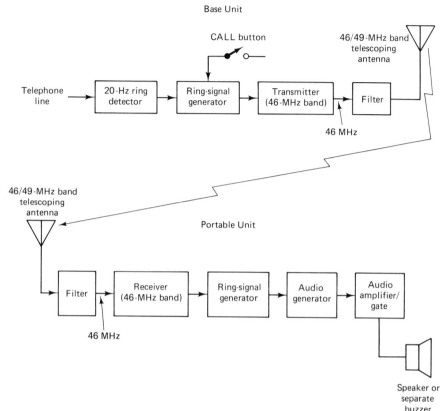

FIGURE 4-3. Basic ring circuits for 46/49-MHz cordless telephones

the 20-Hz ring detector is not used, the check *half-splits* the circuits. (The half-split technique is well known to regular readers of the author's troubleshooting books.)

In this application of the half-split technique, if the telephone rings (with the CALL button pressed), the problem is isolated to the telephone cord or to the 20-Hz ring detector in the base unit. If the telephone does not ring in response to pressing the CALL button, the trouble is probably in the base-unit RF transmitter or the portable-unit receiver.

Now let us make some further checks into the ring circuits before going to the detailed troubleshooting procedures of Chapter 5.

4-3.1 Portable Unit Does Not Ring with CALL Button Pressed

If the portable unit does not ring from the base-unit CALL button, check that the base unit is producing an RF signal.

If the base unit operates in the 1.7-MHz range and you do not have an analyzer, you may have trouble in measuring the base-unit transmitter output. The RF signal must be removed from the a-c power line, demodulated, and then fed to a meter. It is not easy to make up a circuit for such a test. For example, failure to completely remove the 60-Hz a-c component from the RF signal can result in electrical shock, or severe damage to the equipment used in demodulating and measuring the RF signal.

If you have an analyzer to monitor 1.7-MHz base units, simply press the LOAD pushbutton and read the LEVEL meter while holding the base-unit CALL button. The analyzer includes an RF detector circuit which measures the RF level fed into the a-c outlet on the front panel.

If the base unit operates in the 46-MHz range and you are not using an analyzer, connect an RF probe to a multimeter. Touch the probe to the base-unit antenna and read the RF level on the meter while holding the CALL button.

If you have an analyzer to monitor 46-MHz base units, connect an RF demodulator probe to the analyzer and select the DEMOD PROBE mode. Touch the probe to the antenna of the base unit and read the LEVEL meter on the analyzer while holding the CALL button.

For either 1.7- or 46-MHz base units, most cordless telephones should measure about 4 V, but a reading of even 0.5 V permits ringing at close range. If there is no reading, the base-unit transmitter is completely dead. Start by making d-c voltage and resistance measurements in the RF oscillator and output stages of the transmitter circuits, as described in Chapter 5.

If you get a measurable RF output from the base-unit antenna, measure frequency error and ring-signal modulation. If you are not using an analyzer, turn on the base-unit transmitter without ring-signal modulation and measure the RF frequency error with a frequency counter. If the frequency error is more than 2 kHz, the base-unit transmitter frequency should be adjusted. Chapter 5 describes some typical procedures.

If you use an analyzer to check frequency error, connect a cable to the RF IN/OUT jack and clip the leads over the a-c power cord or antenna. Turn on the base-unit transmitter without ring modulation and measure RF frequency error (ΔF). Adjust the base-unit RF transmitter frequency if required.

If the base-unit transmitter is producing an unmodulated signal of the correct frequency and level, check that the transmitter is being modulated by the proper ring signal (typically in the range 700 to 1500 Hz, but check the service literature). Note that the ring signal to the portable unit should be generated with either the CALL button or when a 20-Hz ring signal is detected (by the base unit) on the telephone line.

It is important that the ring signal be of the correct frequency and amplitude. If not, the signal is not passed by the filter in the ring-signal detector of the portable unit. Typically, the ring signal should produce about 3 to 4 kHz of deviation.

If you are not using an analyzer, measure the base-unit transmitter RF output without modulation using a frequency counter. Then press the CALL button, or apply a 20-Hz ring signal to the telephone line, and note the amount of deviation from the unmodulated frequency. If you are using an analyzer, measure the ring-signal frequency deviation by connecting the frequency counter to the SCOPE OUT jack.

If you have an FM deviation meter, you can measure the actual amount of deviation produced by the ring-signal modulation.

If the base-unit transmitter appears to be working (producing both modulated and unmodulated RF signals of correct amplitude and frequency) but the portable unit does not ring, the problem is most likely in the portable-unit receiver. (Of course, if only the ring function is absent and you can hear good voice from the base unit to the portable unit, only the ringer circuits of the portable unit are suspect.)

Set up the analyzer (or an FM-signal generator) to transmit an RF signal to the portable unit. Set the analyzer audio generator (or the audio input to the FM signal generator) to the proper ring frequency, and set the modulation level control (or audio input level to the FM generator) for 4-kHz deviation.

Connect a probe to the output of the discriminator in the portable-unit receiver (Figs. 4-2 and 4-3). If you are using an analyzer, connect the probe output to measure the a-c signal level on the LEVEL meter. This permits the analyzer to function as an audio voltmeter for signal tracing. (If you do not use the analyzer, connect the probe output to an a-c voltmeter.) Either way, it may also be wise to view the signal on an oscilloscope. This will show up any distortion during the demodulation process in the portable-unit receiver.

If you get no signal at the discriminator, the portable-unit receiver circuits (RF and IF) are suspect. Most portable-unit receivers are *dual conversion*, using two local oscillators and two IFs. So check that each of the oscillators is operating, and check each IF signal to make sure that the mixers are operating properly. Refer to Chapter 5 for some typical portable-unit receiver circuits.

If there is an undistorted signal present at the portable-unit discriminator when an RF signal with appropriate FM modulation is applied to the portable unit, the problem is most likely in the ring-signal detector, audio generator, or speaker. (Keep in mind that some portable units have a separate buzzer or ringer rather than using the voice speaker.) Try moving the probe from the discriminator to the output of the ring-signal detector. You should get a deflection on the LEVEL meter (or a-c voltmeter).

If you get an output at the discriminator but not at the output of the ring-signal detector, try varying the frequency of the audio generator used to modulate the RF signal being applied to the portable unit. If this causes a signal at the output of the ring-signal detector, check the frequency where the output appears (or where the portable unit rings). Readjust the base-unit ring-signal generator to that frequency. If there is no output from the ring-signal detector at any audio frequency but you get an output from the discriminator, the ring-signal detector circuits are suspect.

If a signal is present at the output of the ring-signal detector but the portable unit does not ring, move the probe to the output of the audio generator to make sure that an audio tone is being generated. If not, the audio generator circuits are suspect. If the audio generator circuits are producing an audio tone, the speaker (or separate buzzer) is probably defective.

4-3.2 Portable Unit Does Not Ring with Ring Signal Applied

If the portable unit rings normally when the CALL button is pressed but does not ring when 20 Hz is applied to the telephone line input, first make sure that the telephone is actually on-hook.

If you are using an analyzer, check if the VOICE/SIGNAL LEVEL indicator flickers when the RING button is pressed. If not, the analyzer is not generating a ring signal. This is possibly because the analyzer detects an off-hook condition in the base unit. Under these conditions, the analyzer does not generate a ring signal (nor does telephone company equipment). A possible cause for the base unit being off-hook might be another portable unit operating on the same carrier and guardtone (pilot signal) frequencies, and thus "capturing" the base unit. Another possibility is that a fault in the base unit is placing a low impedance across the telephone line.

If you suspect a low-impedance fault or do not have an analyzer, measure the impedance at the telephone cord (with the cord disconnected from the line simulator or telephone line). If the impedance is very high, the telephone is on-hook. If the impedance is low (approximately 250 Ω), the telephone is off-hook. (If the impedance is 0, the cord is probably shorted.)

If the telephone proves to be on-hook, check the 20-Hz ring detector circuit in the base unit. Apply a continuous ring signal and trace the signal through the suspected circuit with an a-c voltmeter. (If the telephone cord is open, there is no ring signal present at the input to the 20-Hz ring detector.)

The ring detector circuit often uses an opto-isolator to switch on the RF transmitter and ring signal generator. Make sure that the opto-isolator is switching. Use a noncontinuous ring signal for this check.

4-4. DIAL-TONE CIRCUIT BASIC TROUBLESHOOTING

Figure 4-4 is the basic troubleshooting tree for dial-tone circuit problems in cordless telephones. Figures 4-5 and 4-6 show the circuits involved for cordless telephones operating in the 1.7/49-MHz and 46/49-MHz ranges, respectively.

As shown in Fig. 4-4, the first step is to place the portable unit in TALK and listen for a dial tone. If you are not connected to a known-good telephone line or using an analyzer, you must simulate a dial tone. This can be done by applying a 0.05-V (rms) audio signal to the line input of the base unit. If you want a true

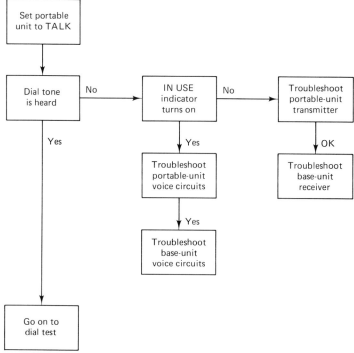

FIGURE 4-4. Basic dial-tone-circuit troubleshooting tree

dial tone, you must use two tones (352 and 440 Hz). However, you can get by with any audio tone between 300 and 3000 Hz.

As shown in Figs. 4-5 and 4-6, the portable-unit RF transmitter and pilot signal (guardtone) generator are turned on when TALK is pressed. The base-unit receiver detects the RF signal and recognizes the guardtone. In turn, the pilot or guardtone detector operates the off-hook relay, which completes the connection to the telephone line and turns on the IN USE indicator. The dial tone on the telephone line is reproduced on the portable unit speaker through the base-to-portable RF link.

If the IN USE indicator turns on, the portable-to-base path is good. That is, it can be assumed that the following circuits are operating normally: portable-unit pilot-signal generator and transmitter, base-unit receiver, pilot-signal detector, and relay driver.

If the IN USE indicator turns on and you hear a dial tone in the portable unit, the base-to-portable path is good. That is, it can be assumed that the following circuits are operating normally: base-unit off-hook relay, hybrid transformer, and transmitter; portable-unit receiver, audio amplifier/gate, and speaker.

Note that the dial-tone test should be performed, no matter what the results of the ring-circuit troubleshooting described in Sec. 4-3. You will see why if you study the circuits involved (Figs. 4-2 through 4-6).

104 *The Basics of Troubleshooting Cordless Telephones*

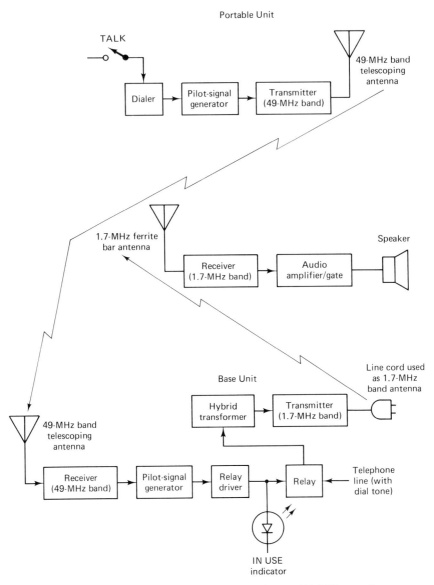

FIGURE 4-5. Basic dial-tone circuits for 1.7/49-MHz cordless telephones

For example, if the portable unit does not ring with either or both the CALL button or 20-Hz ring signal, and the IN USE indicator does not turn on, the base-unit power supply may be defective. This could cause both the RF transmitter and receiver in the base unit to be inoperative.

Dead batteries in the portable unit can cause the same symptom since both the portable transmitter and receiver operate from the same batteries.

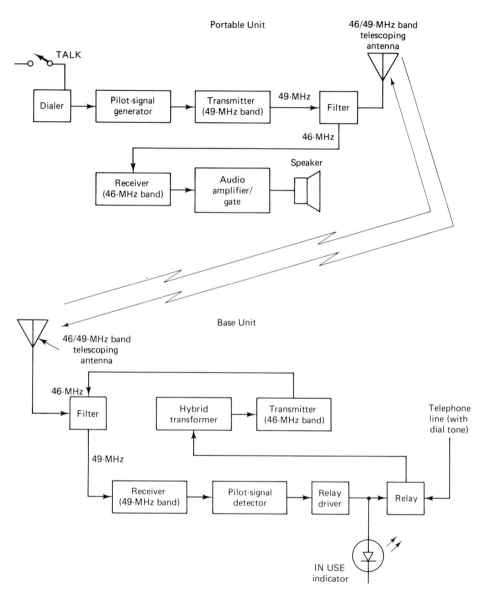

FIGURE 4-6. Basic dial-tone circuits for 46/49-MHz cordless telephones

If the IN USE indicator turns on, and the portable unit rings when the CALL button is used but not when the ring signal is applied, the telephone cord or the connections to the telephone line input of the base unit are probably defective.

If the portable unit does not ring, with either or both the CALL button or 20-Hz ring signal, but the IN USE indicator turns on, listen to the portable unit for static. If you hear static, it is most likely that the base-unit transmitter is not providing a suitable RF carrier. If you hear nothing at the portable unit, the portable-unit receiver is suspect.

4-4.1 No Dial Tone (IN USE Indicator On)

If the dial tone cannot be heard at the portable-unit speaker when the IN USE indicator is on, the problem lies in the base-to-portable path. Since we have already checked most of the circuits involved in the base-to-portable link during the ring test (Sec. 4-3), the most likely defective circuits are the audio amplifier/gate in the portable unit or the voice audio circuits in the base unit.

In portable units where a separate buzzer is used for ringing, the problem can also be in the speaker of the portable unit.

4-4.2 No Dial Tone (IN USE Indicator Fails to Turn On)

If the dial tone cannot be heard at the portable-unit speaker and the IN USE indicator fails to light, the problem is likely to be in the portable-to-base path. Because the portable unit is subjected to harder use, start by checking the portable-unit transmitter.

First check the RF level using a demodulator probe and the LEVEL meter (or a multimeter). As discussed in Sec. 4-3.1, most good cordless telephones should show an RF level of about 4 V, but a reading of even 0.5 V permits close-range operation. The portable unit is completely dead if there is no reading. Start by making d-c voltage and resistance measurements in the RF oscillator and output stages of the portable-unit transmitter circuits, as described in Chapter 5.

If you get a measurable RF output from the portable-unit antenna, measure frequency-error and guardtone (pilot signal) modulation. If the error is more than 2 kHz, the portable-unit transmitter frequency should be adjusted. Chapter 5 describes some typical procedures.

If the portable-unit transmitter is producing a signal of the correct frequency and amplitude, check that the transmitter is being modulated by the proper guardtone or pilot signal. It is important that the signal be of correct frequency and amplitude. If not, the pilot-signal detector in the base unit will not recognize the signal (if the signal is either off-frequency or of low amplitude).

Typically, the pilot signal should produce about 3 to 4 kHz of deviation. Using a frequency counter, the pilot-signal frequency can be measured at the output of a demodulator circuit, such as the demodulator available in an analyzer. (The pilot-signal frequency can be measured on a frequency counter connected to the SCOPE OUT jack of an analyzer.) Typical pilot-signal frequency tolerance is about ±50 kHz. If the pilot frequency is not within specifications, adjust the frequency as described in Chapter 5.

If the portable-unit transmitter appears to be working (producing both RF carrier and pilot modulation of correct amplitude and frequency) but the base-unit

IN USE indicator fails to turn on, the problem is likely to be in the base-unit receiver.

Set up the analyzer (or an FM signal generator) to transmit an RF signal to the base unit. Set the analyzer audio generator (or the audio input to the FM signal generator) to the proper guardtone or pilot frequency, and set the modulation level control (or audio input level to the FM generator) for 3 kHz of deviation at the pilot-signal frequency. Also, apply an audio signal in the range 300 to 3000 Hz with an amplitude high enough to cause an additional 3 kHz of deviation (6 kHz total deviation).

Connect an oscilloscope to the discriminator output of the 49-kHz receiver in the base unit (Figs. 4-5 and 4-6). Check that the guardtone or pilot signal is present. If so, make sure that the signal is of correct frequency. If the pilot signal is absent or is distorted, the base-unit receiver is suspect.

Typically, a base-unit receiver has several stages: an RF amplifier, first IF stage (first fixer, local oscillator, and IF filter), and second IF stage (second mixer, local oscillator, IF filter, and detector). (Base-unit receiver circuits are discussed in Chapter 5.)

First check both local oscillators (first and second IF) to make sure that they are oscillating. If either local oscillator fails to oscillate, you have found the trouble (probably). After the local oscillators are verified as operational, check the output of each stage in the receiver. An RF demodulator probe connected to a multimeter is a convenient method for making such checks. If you are using an analyzer, connect an RF demodulator probe to the DEMODE PROBE jack of the analyzer and read the LEVEL meter.

The next step is to check that the pilot-signal detector in the base unit is detecting the signal and causing the telephone to go off-hook. If the pilot signal is not being detected, vary the frequency of the audio generator (connected to the analyzer or FM generator) to make sure that the pilot-signal detector is not set up to receive a different pilot or guardtone frequency.

If the pilot-signal detector detects a different frequency (from the one that was originally selected), it is easier to change the pilot-signal frequency in the portable unit (to match the frequency of the pilot-signal detector) than to change the filter component values in the pilot-signal detector. Of course, if the portable unit once worked properly with a given base unit, it should not be necessary to change the portable-unit pilot frequency drastically unless there has been some do-it-yourself "experimenting."

4-5. DIAL-CIRCUIT BASIC TROUBLESHOOTING

Figure 4-7 is the basic troubleshooting tree for dial-circuit problems in cordless telephones. Figures 4-8 and 4-9 show the circuits involved for cordless telephones operating in the 1.7/49-MHz and 46/49-MHz ranges, respectively.

If the ring circuits are inoperative (Sec. 4-3) or if you do not get a dial tone

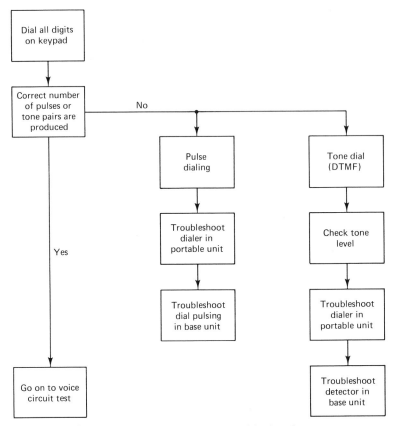

FIGURE 4-7. Basic dial-tone troubleshooting tree

(Sec. 4-4), clear any problems in these circuits before you troubleshoot the dialing circuits.

As shown in Fig. 4-7, the first step is to dial all digits on the keypad and check that the correct number of pulses or tone pairs are produced. If so, the dialing circuits are good and you can go on to the voice-circuit tests of Sec. 4-6. However, if any of the digits produce an incorrect number of pulses or tone pairs, check the dialing circuits as described in the following paragraphs (Sec. 4-5.1 for pulse dial, Sec. 4-5.2 for tone dial).

4-5.1 Pulse-Dial-Circuit Troubleshooting

Most pulse-dial cordless telephones have a dialing IC that turns the pilot signal generator on and off. This causes the base unit to switch on-hook and off-hook, thus pulsing the telephone line. The dialing IC (in the portable unit) is the most likely cause of trouble (assuming that the telephone passes the ring and dial-tone tests, Secs. 4-3 and 4-4) when a cordless telephone fails to produce any digits (as shown on the analyzer display or by a count of the pulses).

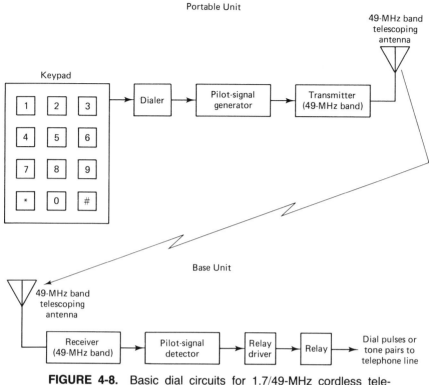

FIGURE 4-8. Basic dial circuits for 1.7/49-MHz cordless telephones

Start by checking for the correct number of pulses from the dialing IC for each digit pressed on the keypad, as described in Sec. 3-4. If one or more digits are absent or abnormal, replace the dialing IC. Of course, make sure that the keypad and wiring between the keypad and dialing IC are good.

The next component to check for dialing problems in cordless telephones is the base-unit relay. Even though the relay operates properly for the ring test and the dial tone test, it is possible that the relay can be defective. For example, if the relay spring loses tension or the relay becomes "sticky," the duty cycle of the pulses is changed and the telephone company equipment (and analyzer) sees the wrong digits. Try replacing the base-unit relay (once you have determined that the dialing IC is good).

4-5.2 Tone-Dial-Circuit Troubleshooting

As with pulse-dial cordless telephones, the components that can cause tone-dialing failure are limited (assuming that the telephone passes the ring and dial-tone tests). The first step is to check the output of the crystal used with the dialing IC. Using an oscilloscope you should be able to see some kind of an

110　The Basics of Troubleshooting Cordless Telephones

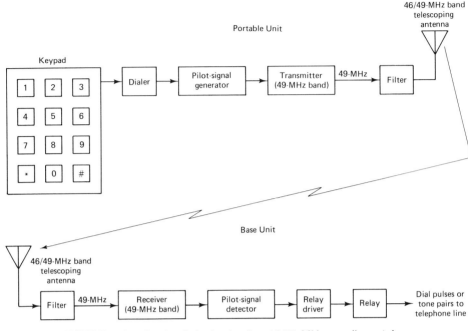

FIGURE 4-9. Basic dial circuits for 46/49-MHz cordless telephones

oscillating trace. Replace the crystal if there is no oscillation. If the crystal is oscillating, check the dialing IC output as described in Sec. 3-4.

The dialing IC should output the proper tone pairs for each digit. If not, replace the dialing IC. If both the dialing IC and crystal are operating, it is possible that there is some attenuation in the signal path. Such attenuation can reduce the tone level to a point where the tone pair is not recognized. If practical, try the portable unit with another base unit.

If the portable unit proves to be operating properly, check the base unit for possible distortion in the detector circuit or for improper level adjustment. With an analyzer, if the VOICE/SIGNAL LEVEL indicator fails to light when any of the keypad digits is pressed, the problem is probably caused by improper level adjustment. If you are not using an analyzer, monitor the base-unit detector with an a-c voltmeter. Typically, the dialing tones should produce an amplitude of at least 0.25 V.

Check the service manual and locate the level-adjustment potentiometer (if any). Adjust the potentiometer for the correct tone level. If the level is correct, the problem is most likely to be caused by distortion in the detector circuit. Connect an oscilloscope to the detector output (or the LINE SCOPE jack if you are using an analyzer) and check the tone-pair signal for distortion. If distortion is present, the detector circuits are suspect.

4-6. VOICE-CIRCUIT BASIC TROUBLESHOOTING

Figure 4-10 is the basic troubleshooting tree for voice-circuit problems in cordless telephones. Figures 4-11 and 4-12 show the circuits involved for cordless telephones operating in the 1.7/49-MHz and 46/49-MHz ranges, respectively.

As shown in Fig. 4-10, the first step is to speak into the portable-unit mouthpiece and listen to the sidetone in the earpiece. Check for adequate voice sidetone level and for good voice quality. If you are using an analyzer, check that the VOICE LEVEL indicator turns on or flickers during voice peaks.

The tests described thus far (ring, dial-tone, and dialing) check out the RF path in both directions and confirm that the telephone cord is good. Also, the

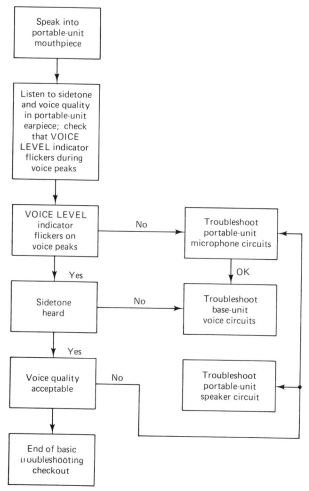

FIGURE 4-10. Basic voice-circuit troubleshooting tree

112   The Basics of Troubleshooting Cordless Telephones

dial-tone test confirms the audio path from the telephone line to the portable unit. The voice tests and procedures described in the following paragraphs also confirm the audio path from the portable unit to the telephone line, as well as rechecking the return audio path for voice level and distortion.

If the VOICE LEVEL indicator does not turn on (even at voice peaks), the voice circuits in the portable-unit transmitter or base-unit receiver are suspect.

Low voice sidetone level may be caused by problems in either direction. That is, the microphone audio is fed through the portable-to-base RF link, coupled by a hybrid transformer to the telephone line, and then back to the earpiece of the portable unit through the base-to-portable RF link. Measuring the voice level at the telephone line half-splits this two-way path to help isolate the cause of attenuation.

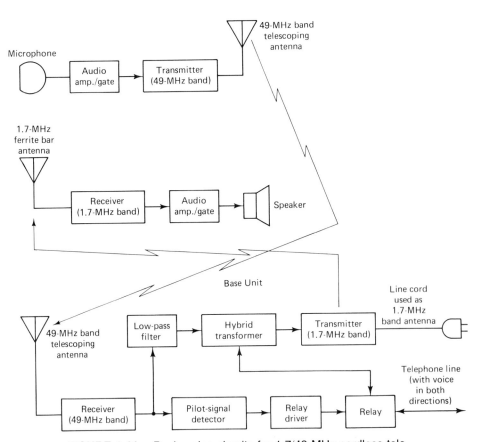

FIGURE 4-11. Basic voice circuits for 1.7/49-MHz cordless telephones

4-6.1 Poor Voice Level

The first step in troubleshooting poor voice level is to check the portable unit. Use an audio generator (or the analyzer AUDIO OUTPUT) to inject an audio tone into the microphone circuit with an a-c voltmeter (or the analyzer LEVEL meter and AUDIO IN functions).

Check that there is a signal being fed to the audio amplifier/gate circuit. The amplifier/gate feeds the amplified voice signal to the transmitter, where the signal is used to modulate the RF carrier. If the voice signal is being fed to the transmitter but the demodulated RF signal (at the base unit) has no audio present, there is probably a broken connection between the audio amplifier/gate and the transmitter.

If the portable unit appears to be operating properly, check the base unit (by substitution if practical). The base-unit receiver demodulates the RF carrier and outputs the guardtone and voice audio signals. The low-pass filter outputs only the voice audio (hopefully) and feeds the voice to the hybrid transformer. In turn, the transformer feeds the audio signal to the telephone line and to the base-unit

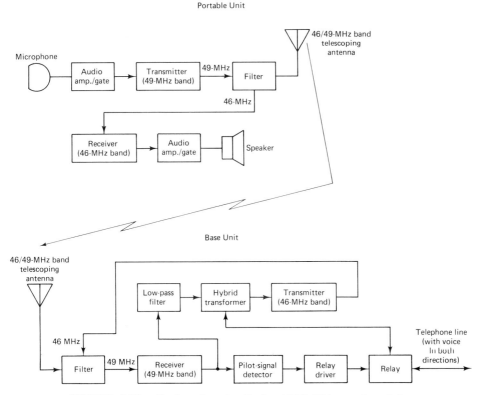

FIGURE 4-12. Basic voice circuits for 46/49-MHz cordless telephones

transmitter. (Note that the signal level fed to the transmitter is much lower than the level sent to the telephone line.) The audio is returned to the portable unit earpiece as sidetone through the base-to-portable RF link.

If the portable unit is good and the system passes all of the other tests (ring, dial-tone, dialing) but the voice level is low, the problem is likely to be in the base-unit low-pass filter or hybrid transformer, or in the connections between the two (since all other base-unit circuits have been proven good).

4-6.2 Low Sidetone Level

If the system has passed all other tests, the most likely cause of low sidetone level is the base-unit hybrid transformer. Check the modulation level of the base-unit transmitter with voice or audio tones applied. Less than 3-kHz deviation at normal voice peak level generally indicates a problem in the hybrid transformer.

4-6.3 Poor Voice Quality

If the voice quality of the cordless telephone proves to be unacceptable, the problem can be in either the portable unit or the base unit. Connect an oscilloscope to the telephone line simulator (or to the LINE SCOPE jack on an analyzer). Inject a low-distortion sine wave at the portable-unit microphone circuit. (The AUDIO OUT jack on the analyzer provides a low-distortion sine wave.)

If distortion is evident on the oscilloscope, use the oscilloscope to examine the audio circuits of the portable-unit microphone circuits and the base-unit receiver discriminator and audio circuit. If no distortion is present at the telephone line simulator, check the hybrid transformer in the base-unit transmitter. Then check the discriminator and speaker circuits of the portable-unit receiver.

4-7. SHORT-RANGE PROBLEMS

The most common complaint about cordless telephones (other than failure of some particular function, such as ringing, dialing, etc.) is limited range between the base unit and the portable unit. Although the most common cause of such limited range is usually low RF power or poor receiver sensitivity, there are other problems to consider before you dive into the circuits. Let us consider some of these problems.

First, all cordless telephones are not created equal! Do not spend hours trying to make a particular cordless telephone perform the same as a telephone of another manufacturer or model. Of course, two identical cordless telephones should perform essentially the same over a given range. Similarly, if a particular system once operated satisfactorily at a given range, the system should continue to do so.

Look for obvious problems such as discharged batteries in the portable unit, antennas not fully extended, and so on. Keep in mind that range is affected by the amount of shielding in the building. Also, on models that use the a-c wiring as the 1.7-MHz antenna, the wiring can act as a shield or attenuation to RF signals in some cases. Finally, short range can be caused by either the base unit or the portable unit, so you must check both. Here, substitution is the most convenient, but not always practical.

Now let us consider the five most likely causes of short range in cordless telephones: Low RF power, poor receiver sensitivity, drifting guardtone or pilot-signal frequency, offset ring signal, and drift in the PLL circuits of both transmitters and receivers.

4-7.1 Low RF Power

The ring and dial-tone tests (Secs. 4-3 and 4-4) measured the *relative RF power* of both base-unit and portable-unit transmitters. As discussed, the typical reading of RF power at the antenna should be about 4 V. (However, there are many cordless telephones that produce far less than 4 V, even under the most ideal conditions, so check the service manual. If you are lucky, the manual may even tell you what RF power to expect!)

Low RF power output can often be corrected by adjustment of the transmitter circuits. (We describe some typical procedures in Chapter 5.) So always try adjustment before you go into any extensive troubleshooting. Also, before any troubleshooting, make sure that you check the RF power, and make adjustments as necessary, on *both the base-unit and portable-unit transmitters*.

4-7.2 Poor Receiver Sensitivity

Testing the sensitivity of a cordless telephone receiver is like testing the sensitivity of any radio receiver. Experience with similar types of cordless telephones will quickly establish normal reference levels for each type. If the receiver has low sensitivity, try correcting the problem with adjustment. Then check the circuits on a state-by-stage basis. Look for defective transistors that do not provide enough amplification, or that produce unwanted attenuation (such as leaking transistors). (We describe some typical procedures for receiver adjustment and test in Chapter 5.)

Using an analyzer and coupling coil (or an FM generator with suitable modulation), it is fairly easy to check receiver sensitivity. Use the analyzer (or FM generator) to transmit a modulated RF signal to the receiver. Check how much attenuation of the RF signal can be selected and clearly receive the signal.

When using the analyzer and a coupling coil directly over the antenna, most cordless telephones should be able to receive the signal with approximately 80 dB of attenuation. (However, some short-range telephones may need lower amounts of attenuation or more RF power. Again, check the service manual.) If the cordless telephone uses an internal antenna (ferrite bar) rather than a telescope antenna, it may be necessary to reduce the attenuation to about 40 dB.

4-7.3 Drifting Guardtone or Pilot-Signal Frequency

Another possible cause of short range is drift of the pilot-signal or guardtone frequency. It is essential that the frequency of the pilot signal generated by the portable unit match the frequency of the pilot-signal detector in the base unit. Use a frequency counter to measure the pilot-signal frequency. If the frequency is off, even by a small amount, range can be reduced drastically and other functions can be affected. Adjust the pilot-signal frequency as necessary. (We describe typical procedures in Chapter 5.)

4-7.4 Offset Ring Signal

As in the case of the pilot signal, it is also essential that the frequency of the ring signal generated by the base unit match the frequency of the ring-signal detector in the portable unit. If not, the ring-signal frequency must be measured and adjusted as required.

4-7.5 Drift in PLL Circuits

In most modern cordless telephones, the frequencies of both the receivers and transmitters are controlled by PLL circuits. Although PLL circuits are crystal controlled, they are subject to drift if the power supply voltage is low (such as when the portable-unit batteries are not fully charged). If you have a short-range problem and you have checked all other functions described in this section, check the frequencies of signals controlled by PLL circuits in both the base unit and the portable unit. Start with the portable-unit transmitter RF output.

4-8. PROBLEMS IN DIGITALLY CODED CORDLESS TELEPHONES

Although it is a "good idea" to use the service manual when troubleshooting any telephone, it is *absolutely essential* that you consult the service manual for the specific model of any digitally coded cordless telephone. Digital coding techniques vary greatly from one manufacturer to another and even from one model to another. (We describe digitally coded telephone troubleshooting in Chapter 5.) The remainder of this chapter provides some background for understanding the Chapter 5 troubleshooting procedures.

Some digitally coded telephones establish a "handshake" before voice communications is possible. In one system, the 1s and 0s of the digital code are created by shifting the RF carrier between two frequencies. The digital code is decoded by the base-unit receiver, which turns on the base-unit transmitter and encoder and sends a code back to the portable unit. The portable-unit receiver decodes the returned digital message and completes the "handshake" (by shutting off the encoder and turning on the voice circuits).

The problem with troubleshooting such a closed-loop system is that almost any condition, either RF or digital, can prevent the handshake from being completed. Generally, when the handshake is not completed after a few trys, such a system shuts down and prevents RF testing. This problem can be overcome by bypassing the digital code. (In most cordless telephones, the digital code can be disabled with jumpers, a switch, or by shorting the digital-code IC output pin to ground.) This makes it possible to test the base unit without the use of coding. In turn, the RF and audio functions can be tested.

Generally, if the base unit works with the digital code bypassed, the encoding or decoding circuits are at fault. Keep several things in mind when testing with the coding disabled. Some digitally coded telephones do not use a guardtone or ring signal. Instead, these telephones send a digital code for ringing and a digital code to signal the base unit to go off-hook. This means that it is very difficult to ring the portable unit without using the base unit. However, all of the other circuits can be tested. If the remaining circuits prove to be good, the problem is most likely to be in the digitally coded ringing system.

One way around this problem is to press the TALK button on the portable unit. This turns on the transmitter of most portable units. You can then monitor the digital code at the output of the base-unit discriminator with an oscilloscope. (In most systems, the digital code is generated at the beginning of transmission.)

Although you probably cannot decipher the actual code, the fact that a code is present and is modulating the portable-unit transmitter proves that half of the circuits are probably good. That is, the portable-unit encoder, as well as the portable-to-base RF link, are both good. The problem is most likely in the base-unit decoder or the base-to-portable RF link.

Note that on some telephones the transmitter may shut off after sending the digital code a few times if no acknowledgment is returned. So it may be necessary to turn the TALK switch off, and then back on to resume testing.

Some cordless telephones have built-in systems for testing the equipment. One such method involves setting all of the digital-code selection switches to the ON position. This disables certain signals or tones that are generated for calling the user's attention to a problem. (You can see why it is so essential that you have the service manual for digitally coded telephones!)

Another way of testing the base-unit transmitter (if it is not practical with jumpers or switches) is to apply a ring signal to the telephone line (or at the line input to the base unit). This should turn on the base-unit transmitter. With the base-unit transmitter turned on and generating a ring code, the portable-unit receiver can be checked by monitoring the discriminator output or the "code-in" input to the portable-unit digital encoder/decoder IC.

Also keep in mind that with some digital coding system, pulse dialing or tone dialing is done digitally between the portable unit and the base unit. The base-unit decoder then generates the actual pulsing or tone pairs and applies the signals to the telephone line.

Finally, some manufacturers use a *dual guardtone* system to improve security, without the complexity of digital coding. In one such system, a 700-Hz tone is sent initially, then a 6-kHz tone. Both tones must be of correct frequency, and must occur within a certain time window, to capture the telephone line.

The analyzer (or other combination of FM test equipment) could be used to simulate this dual-guardtone system for testing. For example, by using both an external audio generator and the internal variable generator of the analyzer, and then switching from external to internal modulation modes at the proper time, you might be able to produce the correct combination of timed signals. However, it is not an easy trick. Why not try using the particular cordless telephone you are servicing?

5

CORDLESS TELEPHONE TROUBLESHOOTING

This chapter describes a series of troubleshooting and service notes for some typical cordless telephones. As discussed in the Preface, it is not practical to provide a specific troubleshooting procedure for every telephone. Instead, we describe a universal troubleshooting approach in Chapter 4 and then follow up with specific examples in this chapter.

The chapter is divided into three parts: circuit descriptions, test/adjustment procedures, and circuit-by-circuit troubleshooting.

By studying the circuits found in this chapter, you should have no difficulty in understanding the schematic and block diagrams of similar telephones. This understanding is essential for logical troubleshooting and service, no matter what type of electronic equipment is involved. No attempt is made to duplicate the full schematics for all circuits. Such schematics are found in the service literature for the particular telephone. Instead of a full schematic, the circuit descriptions are supplemented with partial schematics and block diagrams that show such important areas as signal flow paths, input/output, adjustment controls, test points, and power-source connections. These are the areas most important in service and troubleshooting. By reducing the schematics to these areas, you will find the circuit easier to understand and will be able to relate circuit operation to the corresponding circuit for the telephone you are servicing.

Because adjustments are closely related to troubleshooting, we describe typical adjustment procedures for cordless telephones. Keep in mind that these specific procedures apply directly to the circuits described in this chapter. When servicing other telephones, *you must follow* manufacturer's service instructions exactly. Each type of telephone has its own adjustment points and procedures, which may or may not be different from the procedures of other telephones.

Using the adjustment procedure examples, you should be able to relate the procedures to a similar set of adjustment points on most cordless telephones. Where it is not obvious, we also describe the purpose of the adjustment procedure. The signals measured at various test points during adjustment are also included here. By studying these signals, you should be able to identify typical signals found in most telephones, even though the signals may appear at different points for your particular unit.

The circuit-by-circuit troubleshooting approach is based on trouble symptoms. A series of trouble symptoms is listed at the beginning of the troubleshooting section (Sec. 5-6). These symptoms could apply to any cordless telephone but are related specifically to the telephone circuits described in this chapter. Each trouble symptom is referred to a specific section (Secs. 5-7 through 5-28). These sections show the circuits most likely to be involved in such trouble symptoms, and describe the specific components to be checked as well as the logical order for making the checks. The sections also make reference to any test/adjustment procedures that could affect the specific involved for a particular symptom.

5-1. INTRODUCTION TO CORDLESS TELEPHONE CIRCUITS

The majority of the circuits described in this chapter are part of the Cobra Model CP-440S or the Cobra Model CP-445S, shown in Figs. 5-1 and 5-2, respectively. It is assumed that you have already studied the basic operating controls, operating procedures, and overall functions of similar units, as described in Chapters 1 through 4.

Each unit is a cordless telephone which operates similar to a standard telephone except that communication between the portable unit (also called the *remote* or *handset* in some literature) and the base unit is wireless.

The base unit is powered by a 120-V, 60-Hz source to operate the FM transmitter and receiver. The portable unit FM transmitter and receiver are powered by a 3.6-V nickel–cadmium rechargeable battery.

The telephone cord from the base unit is connected directly to a standard telephone jack and communicates with the central-office telephone lines in the same way as a conventional telephone.

The portable-unit transmitting frequency band is between 49.670 and 49.990 MHz. This band is allocated 10 RF channels, each of which is assigned to a specific crystal-controlled RF carrier frequency (to reduce interference).

The base-unit transmitting frequency band is between 46.610 and 46.970 MHz. This band is also allocated 10 RF channels, each assigned a specific crystal-controlled RF frequency.

The portable unit receiver is highly sensitive, with a crystal-controlled local oscillator and a narrowband IF to receive the base-unit transmitted carrier frequency. Similarly, the base-unit FM receiver receives the portable-unit transmitted carrier frequency to form the communications link.

Introduction to Cordless Telephone Circuits 121

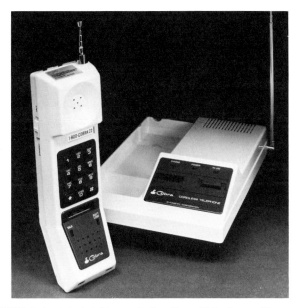

FIGURE 5-1. Model CP-440S Cordless Telephone (Courtesy of Dynascan Corporation)

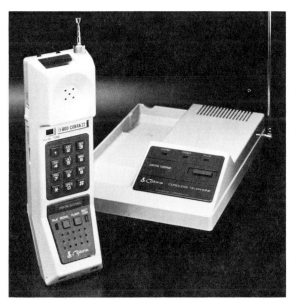

FIGURE 5-2. Model CP-445S Cordless Telephone (Courtesy of Dynascan Corporation)

The operational system for both units includes selectable, digitally coded signal security to prevent unauthorized use of the telephone. Both units have 32 possible digital codes and are capable of call-waiting functions. The Model CP-440S uses pulse dialing. Either pulse dialing or tone dialing can be selected on the Model CP-445S.

5-2. BASE-UNIT INCOMING COMMUNICATIONS CIRCUITS

Figure 5-3 is a block diagram of the base-unit circuits involved in the incoming communications link. Once the base unit is connected into the 120-V, 60-Hz source, the receiver section turns on and the transmitter output section remains in the standby state.

5-2.1 Base-Unit Receiver RF Circuits

Figure 5-4 shows the base-unit receiver circuits. When the desired portable-unit transmitted RF signal is received by the base-unit telescoping antenna, the RF signal is first amplified by Q1 and then fed to the first mixer/IF stage (Q2), which is crystal controlled (X1) by local oscillator Q5 to produce 10.695 MHz (the first IF signal frequency).

The first IF signal is filtered through CF1 and fed to pin 16 of IC1, where the signal is converted by a second mixer (also crystal controlled, X3) to produce 45 kHz (the second IF signal frequency). The second IF signal is filtered by CF2 and fed back into IC1 for limiting and demodulation. The demodulated output appears at pin 9 of IC1.

Figure 5-5 shows the internal circuits of IC1. Keep in mind that the IC circuits are not accessible for test, adjustment, or troubleshooting and must be checked on an input/output basis (as is the case with all ICs).

5-2.2 Base-Unit Pilot-Tone and Security-Code Circuits

Figure 5-6 shows the base-unit pilot-tone and security-code circuits. The 5.3-kHz pilot tone (guardtone) at IC1-9 is detected first and then fed through VR2 to tone decoder IC2 at input terminal 3. Figure 5-7 shows the internal circuits of IC2. The frequency selected by IC2 is determined by the setting of VR4 connected at pins 5 and 6. The output of IC2 is a low at pin 8, when the incoming pilot tone is at the same frequency as that selected by VR4. The low at IC2-8 is applied to reset the IC4 flip-flop (FF) at pin 1.

The two-word security-coded signal at IC1-9 is applied through Q11 back to pin 10 of IC1. After processing, the signal is applied to decoder IC202 at pin 9. If the received coded information matches the code settings of switch S202, the output at pin 11 of IC202 goes low.

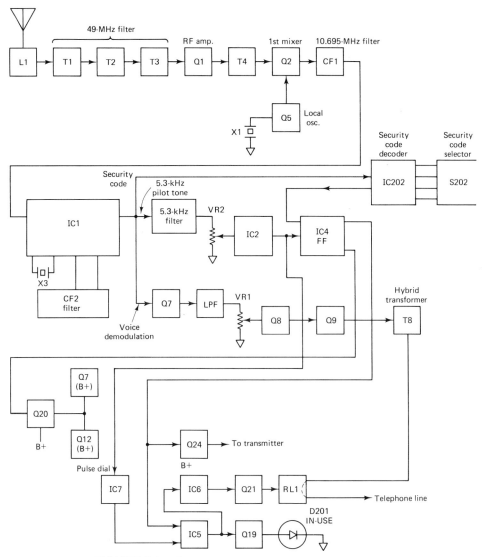

FIGURE 5-3. Base-unit incoming communications link

The low output form IC202-11 is inverted by IC7 to set the pin-3 output of IC4 flip-flop high (Q) and the pin-11 output low ($\bar{Q}$). The pin-3 output of IC4 turns on the transmitter B+(Q24), drives NAND gate IC5 pin 10 low to turn on Q19 (the IN-USE indicator D201 driver), and drives inverted IC6 pin-8 high to turn on hook relay RL1 (through Q21). This connects the telephone line to the base-unit hybrid transformer.

The pin-11 output of IC4 turns on Q20 to supply d-c voltages to audio preamplifier Q7 and audio-input modulation amplifier Q12.

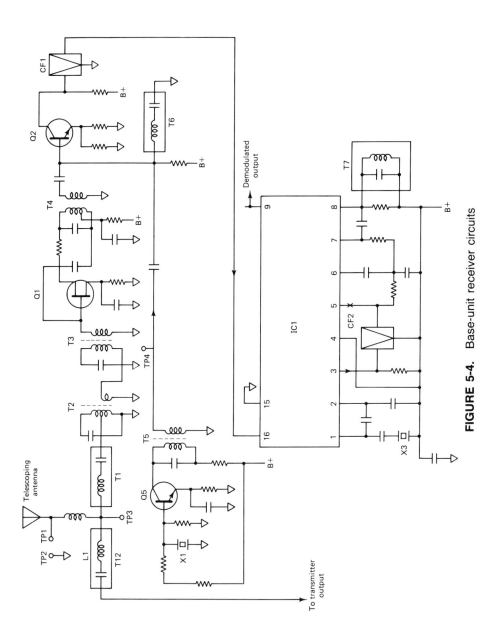

FIGURE 5-4. Base-unit receiver circuits

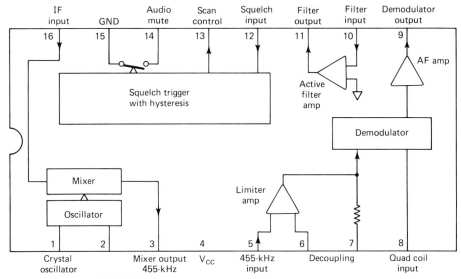

FIGURE 5-5. Internal circuits of IC1 (FM IF)

5-2.3 Base-Unit Voice Demodulation Circuits

Figure 5-8 shows the base-unit voice demodulation circuits. The voice signal from the portable unit is demodulated by IC1 and is fed through Q7 to a low-pass filter (C22, C23, C24, L2). The filter has a cutoff frequency of 3.5 kHz to eliminate the pilot signal and any harmonic distortion components present on the telephone line. Level control VR1 is adjusted to set the proper audio output level of the signal applied through amplifiers Q8/Q9 and the hybrid transformer T8 to the telephone line. Z1 protects the circuit from excessive current surges or "spikes."

5-2.4 Pulse-Dial Processing Circuits

Figure 5-9 shows the base-unit pulse-dial processing circuits. Dial signals or pulses sent from the portable unit are produced by turning the 5.3-kHz pilot signal on and off. This signal is fed from IC1-9 to the input of tone decoder IC2 at pin 3 and produces dial pulses (square waves) at the output of IC2-8. The square-wave dial pulses are applied to hook relay RL1 through IC7, IC5, IC6, and Q21. RL1 is switched on and off by the square waves and thus pulse-dials the telephone line.

5-3. BASE-UNIT OUTGOING COMMUNICATIONS CIRCUITS

Figure 5-10 is a block diagram of the base-unit circuits involved in the outgoing communications link.

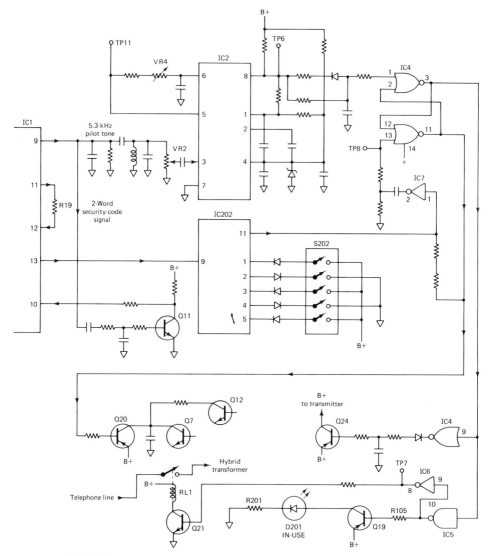

FIGURE 5-6. Base-unit pilot-tone and security-code circuits

5-3.1 Base-Unit Telephone Line Signal Processing Circuits

Figure 5-11 shows the base-unit telephone line circuits. The incoming voice or other signals on the telephone line are coupled by hybrid transformer T8 and fed to modulation amplifier Q12. Z1 protects the circuits from excessive voltage spikes. VR3 sets the amplitude of the telephone line signals. The signal from Q12 is applied to varicap diode D3 and the associated crystal-controlled FM modulator

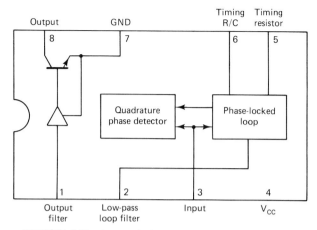

FIGURE 5-7. Internal circuits of IC2 (tone decoder)

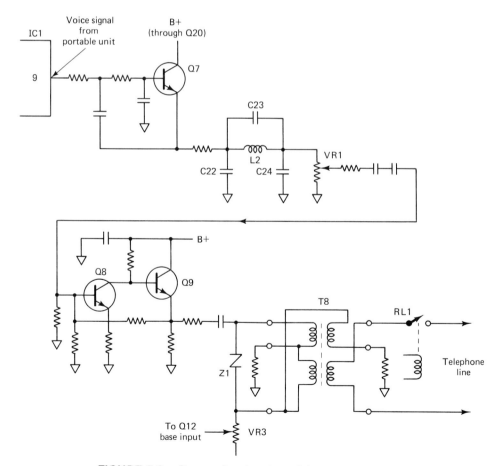

FIGURE 5-8. Base-unit voice demodulation circuits

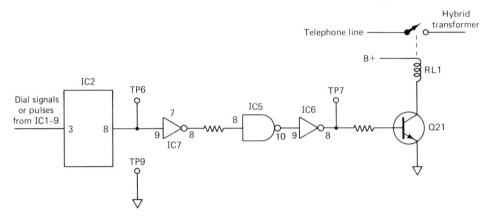

FIGURE 5-9. Base-unit pulse-dial processing circuits

circuit of Q16. L4 and TC2 set the channel frequency, which is controlled by crystal X5.

The fundamental FM signal is tripled and amplified by Q17/Q18 and fed to the telescoping antenna through an impedance-matching network (T11/T12). L1 is the antenna loading coil. T9, T10, T11, and T12 are aligned for maximum RF output to the antenna.

5-3.2 Base-Unit Ring Detection and Transmission Turn-On Circuits

Figure 5-12 shows the base-unit ring detection and transmission turn-on circuits. The ring signal from the telephone line incoming call is detected by a neon coupler OC1. The output from OC1 is connected to IC4-6 and IC5-13. OC1 goes low when the telephone line ring voltage is applied.

The output of IC4-4 goes high and turns on a 20-Hz oscillator (IC6, pins 2 and 10) which switches a 2.1-kHz oscillator (IC6, pins 6 and 12) on and off. Transistor Q14 is normally turned on by the B+ through D11 and R47. This shorts the output of the 2.1-kHz oscillator to ground. When the 20-kHz oscillator is turned on, Q14 is turned on and off at a 20-kHz rate, permitting the 2.1-kHz oscillations to pass. The 2.1-kHz signal is applied to the modulator Q16 for transmission to the portable unit.

The transmitter is turned on by the high output at IC5-11 (through Q24). The code generator IC201, and the IN-USE indicator D201 driver Q19, are turned on by the low output at IC6-4.

With the transmitter on, and the signal being picked up by the portable unit, the 2.1-kHz oscillator (interrupted by the 20-Hz signal) simulates the incoming telephone line ring signal, while the code generator IC201 modulates the carrier with the desired code (determined by the settings on S202).

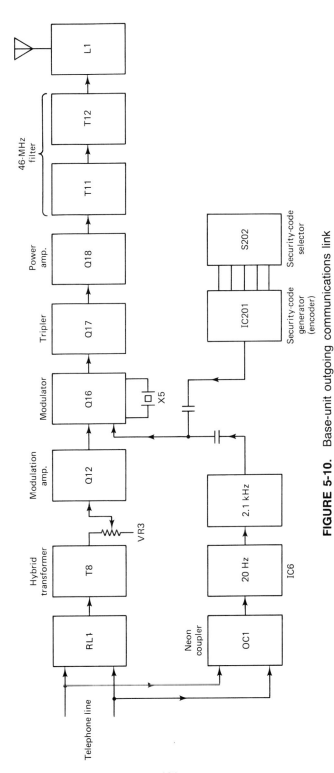

FIGURE 5-10. Base-unit outgoing communications link

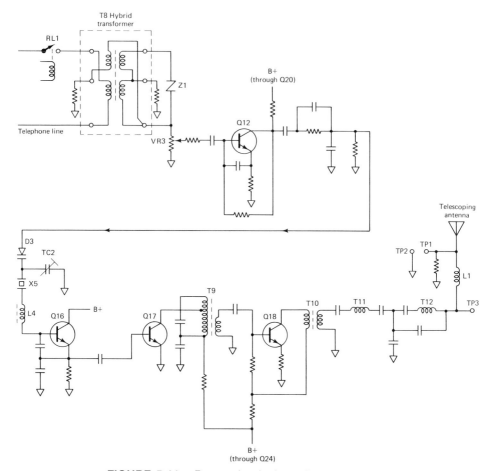

FIGURE 5-11. Base-unit telephone line circuits

5-3.3 Base-Unit CALL Button Circuits

Figure 5-13 shows the base-unit CALL button circuits. When the CALL button S201 is pressed, IC5-12 goes low. This turns off Q14 (through D11 and R47) and turns on the 2.1-kHz oscillator (through IC6 and IC5). The transmitter is also turned on (because of the high at IC5-11), as is the code generator IC201 (because of the low at IC6-4).

With the transmitter on, the 2.1-kHz oscillator modulates the RF carrier to simulate a ring signal, while the code generator IC201 modulates the carrier with the correct digitally selected code (determined by the settings of S202).

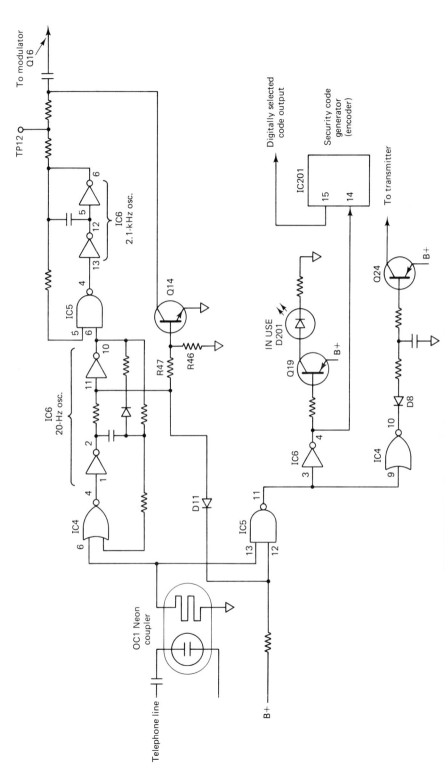

FIGURE 5-12. Base-unit ring detection and transmission turn-on circuits

132 Cordless Telephone Troubleshooting

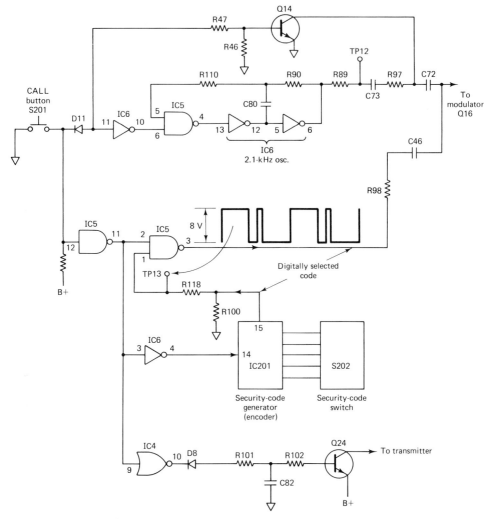

FIGURE 5-13. Base-unit CALL button circuits

5-3.4 Base-Unit Power Supply Circuits

Figure 5-14 shows the base-unit power supply circuits. The power supply consists of step-down transformer T13, a bridge rectifier D15, and a regulator circuit Q26. All of the power supply circuits are conventional and do not require explanation. (If they do require explanation, you may have considerable trouble servicing any cordless telephone!)

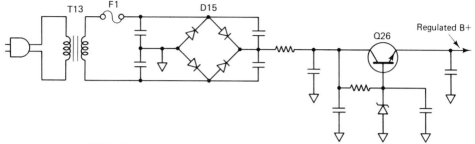

FIGURE 5-14. Base-unit power supply circuits

5-4. PORTABLE-UNIT COMMUNICATIONS CIRCUITS

Figure 5-15 is a block diagram of the portable-unit circuits involved in the communications link.

5-4.1 Portable-Unit Receiver RF Circuits

Figure 5-16 shows the portable-unit receiver and voice circuits. When the desired 46-MHz channel RF signal is received by the remote telescoping antenna, the signal is first amplified by Q301 and then fed to the first mixer/IF(Q304), which is crystal controlled (X301) by local oscillator Q302 to produce 10.695 MHz (the first IF signal frequency).

The first IF signal is filtered through CF301 and fed to pin 16 of IC301, where the signal is converted by a second mixer (also crystal controlled by X303) to produce 455 kHz (the second IF signal frequency). The second IF is filtered by CF302 and fed back into IC301 for limiting and demodulation. The demodulated output appears at pin 9 of IC301.

Figure 5-17 shows the internal circuits of IC301. The voice output at pin 9 of IC301 is applied to the earpiece speaker SP401 through amplifiers Q402, Q403, and Q404. The amplitude of the voice signal can be controlled by HIGH/LOW switch S303, which cuts series resistance R314 in or out of the circuit. S301 must be in TALK and S302 must be ON for the voice circuits to turn on.

5-4.2 Portable-Unit Modulation and Transmitter RF Circuits

Figure 5-18 shows the portable-unit modulation and transmitter RF circuits. Voice signals picked up by the microphone are amplified by Q305/Q306 and fed to the transmitter modulator. The signal from Q305 is applied to varicap diode D302 and the associated crystal-controlled FM modulator circuit of Q311. L303 and TC302 set the channel frequency, which is controlled by crystal X305.

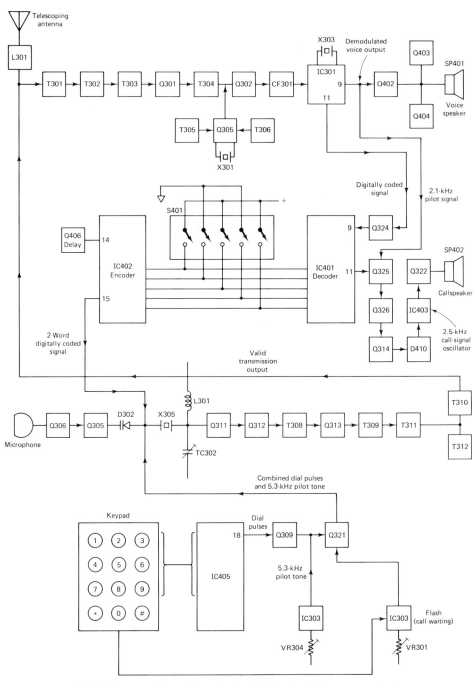

FIGURE 5-15. Portable-unit circuits involved in communications link

Portable–Unit Communications Circuits 135

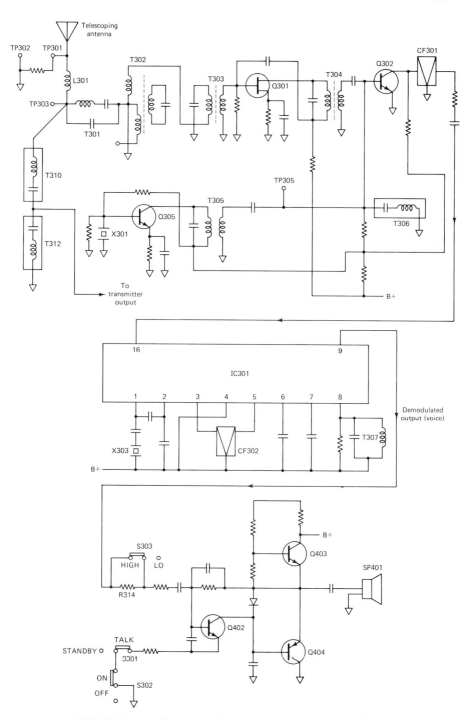

FIGURE 5-16. Portable-unit receiver and voice circuits

136   Cordless Telephone Troubleshooting

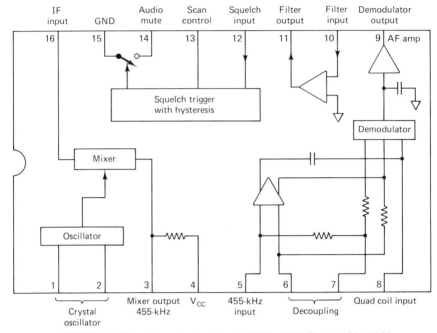

FIGURE 5-17. Internal circuits of IC301 (FM IF, portable unit)

The fundamental FM signal is tripled and amplified by Q312/Q313 and fed to the telescoping antenna through an impedance-matching network. L301 is the antenna loading coil. T308 through T312 are aligned for maximum RF output to the antenna.

5-4.3 Portable-Unit Tone-Dial Circuits

Figure 5-19 shows the portable-unit tone-dial circuits. Tone pairs are generated within IC404, which is controlled by 3.759545-MHz crystal X401. Figure 5-20 shows the internal circuits of IC404. When PULSE/TONE switch S402 is in the TONE position and a keypad button is pressed, the corresponding tone pairs are produced by IC404 at pin 16. The tone pairs are applied to the modulation varicap diode D302 for transmission to the base unit and telephone line. The tone pairs are also applied to the earpiece speaker SP401 (as the *keystroke tone*) through audio amplifiers Q402, Q403, and Q404.

5-4.4 Portable-Unit Call-Signal Circuits

Figure 5-21 shows the portable-unit call-signal circuits. To receive a call from the base unit, the portable-unit TALK/STANDBY switch S301 must be in the STANDBY position. The digitally coded signal and the 2.1-1Hz pilot signal must be received simultaneously (valid transmission) to produce a call tone.

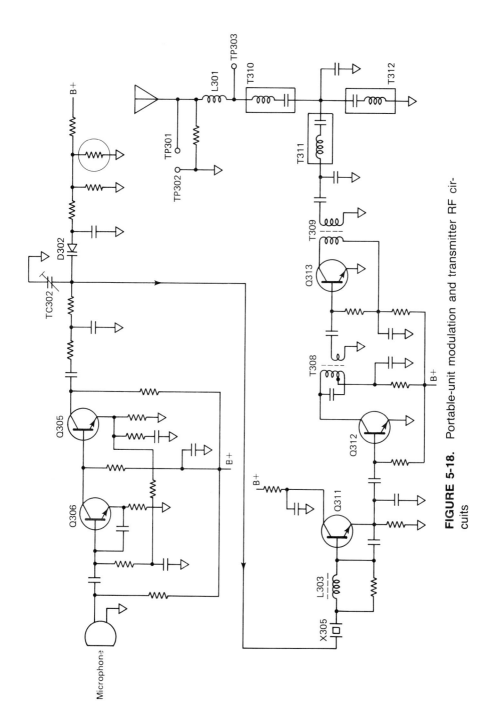

FIGURE 5-18. Portable-unit modulation and transmitter RF circuits

138   Cordless Telephone Troubleshooting

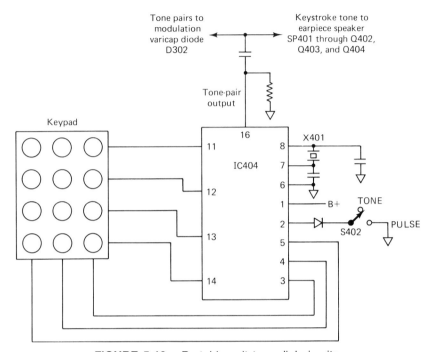

FIGURE 5-19. Portable-unit tone-dial circuits

The valid transmission output of decoder IC401 at pin 11 goes high when the proper code (matching that selected by S401) from the base unit is applied to pin 9 of IC401. The pin-9 input of IC401 receives the digitally coded signal from IC301-11 through Q324. Figure 5-22 shows the internal circuits of IC401.

The high at IC401-11 turns on the 2.1-kHz pilot tone amplifier Q325. The 2.1-kHz pilot tone (from the base unit through IC301) is rectified by Q326 to turn on Q314. Diode D410 is reverse biased when Q314 turns on. This activates the 2.5-kHz call-signal oscillator (composed of inverters in IC403). The 2.5-kHz signal is then coupled to amplifier Q322, which drives the call speaker SP402.

When IC401-11 is low (no valid transmission), D305 is forward biased. This shorts the base of Q322 to ground and prevents the 2.5-kHz call signal from passing to SP402.

5-4.5 Portable-Unit Dial-Signal Circuits

Figure 5-23 shows the portable-unit dial-signal circuits. At the instant TALK/ STANDBY switch S301 is placed in the TALK position, the transmitter and the 5.3-kHz pilot-tone generator (composed of the inverters in IC303) are turned on, and transmission begins. After a short delay (produced by the Q406) the code generator IC402 pin 14 goes low, producing a two-word digitally coded signal at pin 15. Figure 5-24 shows the internal circuits of IC402. The code signal is applied to the

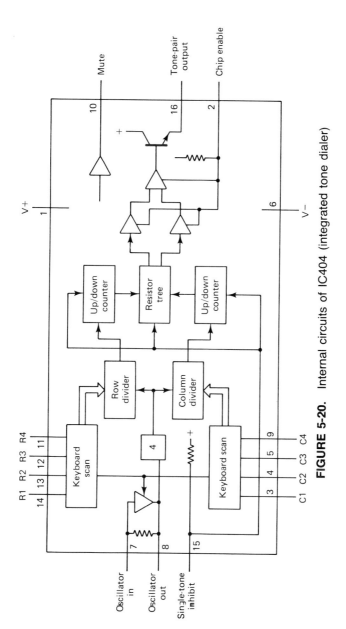

FIGURE 5-20. Internal circuits of IC404 (integrated tone dialer)

139

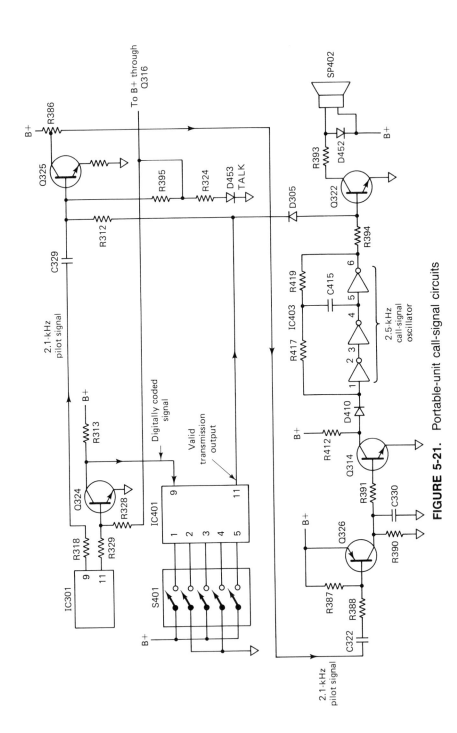

FIGURE 5-21. Portable-unit call-signal circuits

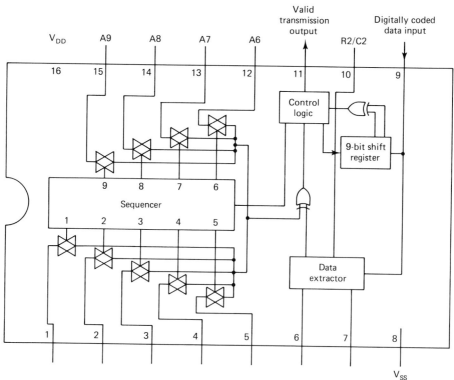

FIGURE 5-22. Internal circuits of IC401 (decoder)

modulator through C424, R406, and R337. At this point, the portable unit and the base unit are in direct communication with each other.

When any number on the keypad is pressed, the dialing pulse generator in IC405 generates corresponding dial pulses at output pin 18. Figure 5-25 shows the internal circuits of IC405. The pulses at pin 18 are applied through Q309 to turn the 5.3-kHz pilot-tone generator output transistor Q321 on and off at the dial-pulse rate. Since the pilot-tone output is connected to the modulator varicap D302, the transmitter modulation varies at the dial-pulse rate. The frequency of the pilot tone is adjusted to 5.3 kHz by VR304.

5-4.6 Portable-Unit Flash-Operation/Call-Waiting Circuits

Figure 5-26 shows the portable-unit flash-operation/call-waiting circuits. The FLASH button is used only with call-waiting service (available from many telephone companies). With call-waiting, you press and *quickly release* the FLASH button to put a caller on "hold" and take a second call on the same line.

On many cordless telephones, if you hold the FLASH button *for a few seconds*, the pilot tone is removed, the telephone line is disconnected, and the

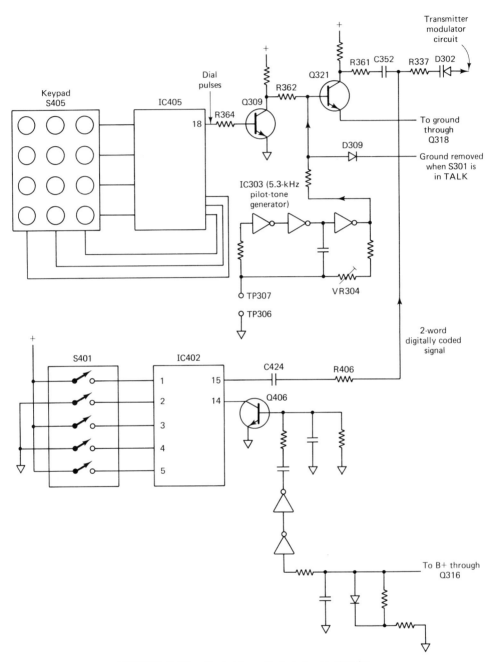

FIGURE 5-23. Portable-unit dial-signal circuits

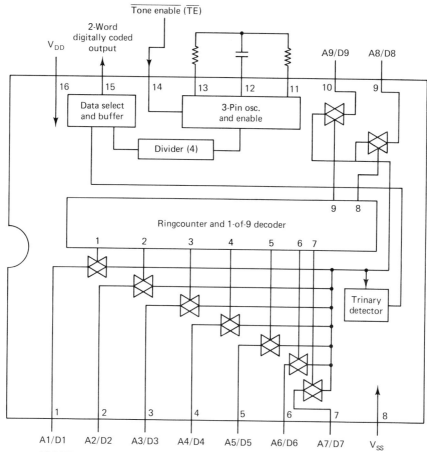

FIGURE 5-24. Internal circuits of IC402 (code generator or encoder)

communications link between the portable unit and base unit is removed. In effect, you "hang up" without returning the portable unit to the base unit.

When the FLASH (#) button on the keypad is pressed, IC303 pin 4 goes high and pin 2 goes low, discharging C353. This causes D308 to conduct and turns off the 5.3-kHz pilot tone for approximately 500 ms (the time required for C353 to charge). Control VR301 is adjusted for the correct time interval.

Once C353 is charged, the 5.3-kHz pilot tone is restored. This temporary interruption of the pilot tone causes a temporary release of the on-hook relay in the base unit (relay RL1 in Fig. 5-6). In turn, this opens the telephone line (momentarily) and signals the telephone company that you are ready for any waiting calls.

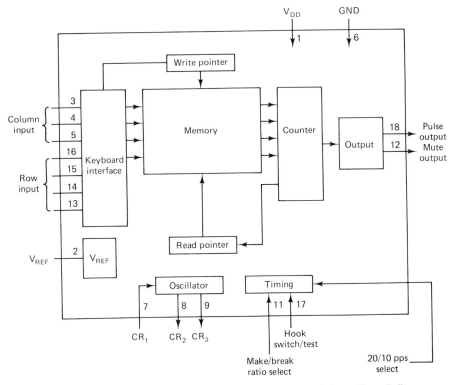

FIGURE 5-25. Internal circuits of IC405 (pulse dialer with redial)

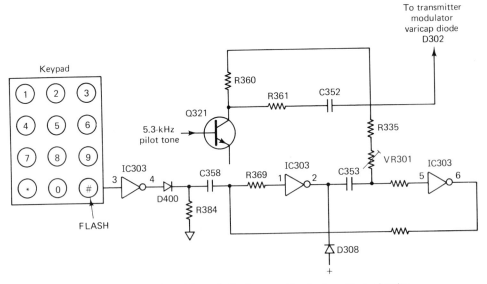

FIGURE 5-26. Portable-unit flash operation/call-waiting circuits

5-5. ADJUSTMENT PROCEDURES

The following paragraphs describe complete adjustment procedures for the cordless telephones described in Sec. 5-2 through 5-4. The electrical locations for the adjustment controls and measurement points (test points, or TP) are given in the diagrams of Figs. 5-3 through 5-26. The following paragraphs refer to the illustrations that show the adjustment and measurement points.

Keep in mind that the procedures described here are the only procedures recommended by the manufacturer for that particular model of cordless telephone. Other manufacturers may recommend more or less adjustment. It is your job to use the correct procedures for each player you are servicing.

Also remember that some disassembly and reassembly may be required to reach test and/or adjustment points. We do not include any disassembly/reassembly for two reasons. First, such procedures are unique and can apply to only one model of telephone. More important, disassembly and reassembly are areas where telephone service literature is generally well written and illustrated. Just make sure that you observe all the notes, cautions, and warnings found in the disassembly/reassembly sections of the telephone service literature.

5-5.1 Test Equipment Required

The following test equipment is recommended by the manufacturer.

1. Sinnader Model S103, or equivalent
2. Wavetek Automatic Modulation Meter, Model 4101 or equivalent
3. Oscilloscope
4. RF voltmeter (high-impedance input)
5. FM signal generator, 45 to 50 MHz, 50-Ω output impedance
6. Audio generator, 300 to 8000 Hz
7. AC VTVM
8. DVM
9. Frequency counter, 50 MHz or higher
10. 600-Ω resistor
11. 8-Ω, 0.5-W resistor

5-5.2 Base-Unit Transmitter Alignment

Preparation for alignment. Disconnect the rod antenna. Connect all of the test equipment (including the resistive network) to the base unit as shown in Fig. 5-27. Terminate the telephone line input with a 600-Ω resistor. Connect the power cord to a 120-V, 60-Hz outlet. Connect a jumper between TP6 and TP9

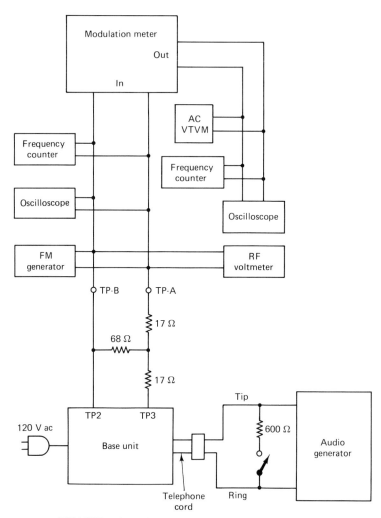

FIGURE 5-27. Base-unit transmitter alignment

(Fig. 5-9) to actuate telephone line relay RL1. Using a screwdriver or jumper, momentarily short pins 13 and 14 of IC4 (Fig. 5-6). This actuates the IC4 flip-flop and turns on the transmitter. Allow at least 5 minutes for warm-up of the base-unit circuits. Set the filter and deemphasis switches on the modulation meter to off.

Transmitter power output. Align TP9, TP10, TP11, and TP12 (Fig. 5-11) for maximum indication on the RF voltmeter. A calibrated RF voltmeter should read between 150 and 250 mV(rms) at TP-A and TP-B of the resistive network.

Transmitter frequency. The frequency counter should read within ±500 Hz of the designated channel frequency. (Figure 2-9 shows channel frequencies for cordless telephones.) If the transmitter frequency is not within tolerance, adjust L4 and TC2 (Fig. 5-11) as necessary.

Call signal deviation and frequency. Connect TP13 to ground (Fig. 5-13) to disable the security-code circuits, and thus remove the security-code signal from the carrier. Press the CALL button and observe the modulation meter for a deviation reading between ±2.5 and ±3.7 kHz. If necessary, adjust L4 and TC2 (Fig. 5-11) for correct deviation. Simultaneously check the frequency counter for a carrier frequency reading of −100 ±20 Hz. If it is necessary to adjust L4 and TC2, repeat the transmitter power output alignment. Connect the frequency counter to the output of the modulation meter. The call signal frequency should read 2100 ± 200 Hz. Remove the jumper from TP13.

Coded signal deviation. Connect pin 10 of IC6 (Fig. 5-13) to ground to disable the 2.1-kHz call signal oscillator. Press the CALL button and observe the modulation meter for a deviation reading between ±4 and ±6 kHz. Note that the modulation meter cannot indicate a correct reading because of the square-wave input signal (that shown at TP13 in Fig. 5-13). The actual deviation is approximately half the modulation meter reading.

To check the IC201 security code generator (encoder) output, connect an oscilloscope d-c input to TP13. Adjust the oscilloscope vertical sensitivity to 2 V/div and sweep-time division to 5 ms. Press the CALL button and observe that the code signal display is similar to that shown in Fig. 5-13.

Modulation sensitivity. Set the FM generator RF output to 2 mV, without modulation. Connect the audio generator set at 1 kHz across the 600-Ω resistor. Adjust the audio generator output level until the modulation meter indicates ±2 kHz deviation. The audio generator output should read −15 ± 2 dBm. Adjust VR3 (Fig. 5-11) if necessary for correct deviation.

5-5.3 Base-Unit Receiver Alignment

Preparation for alignment. Disconnect the rod antenna. Connect all of the test equipment (including the resistive network) to the base unit as shown in Fig. 5-28. Terminate the telephone line input with a 600-Ω resistor. Set the FM generator modulation frequency to 1 kHz with ±2 kHz deviation. Connect a jumper between TP6 and TP9 (Fig. 5-9) to actuate RL1.

Local oscillator adjustment. Connect the frequency counter to TP4 (Fig. 5-4). Adjust T5 for correct crystal frequency readout.

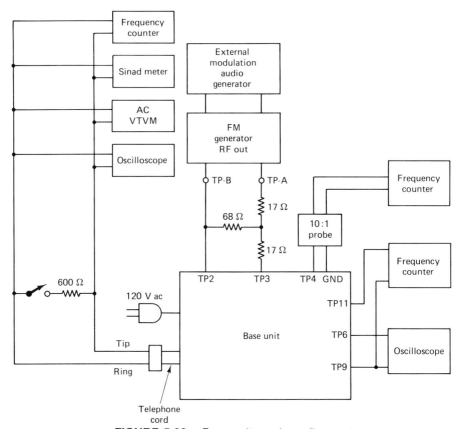

FIGURE 5-28. Base-unit receiver alignment

IF alignment. Turn on the transmitter circuits by momentarily shorting pins 13 and 14 of IC4 (Fig. 5-6). Adjust the FM generator RF output for 1 mV. Adjust T7 (Fig. 5-4) for maximum audio output on the AC VTVM.

Sensitivity adjustment. Adjust the FM generator RF output to obtain 12 dB indication on the Sinad meter. Align T1, T2, T3, and T4 (Fig. 5-4) for minimum reading on the Sinad meter. During alignment, keep reducing generator output to maintain 12-dB indication on the Sinad meter. With T1 through T4 adjusted for minimum, adjust T12 for minimum indication on the Sinad meter while observing the RF meter to maintain an optimum RF level output.

Telephone line output level. Set the FM generator RF output for 2 mV. Adjust VR1 (Fig. 5-8) for -8 dBm output.

Pilot frequency adjustment. Set the FM generator RF output to 2 mV without modulation. Connect the frequency counter to TP11 (Fig. 5-6). Adjust

VR4 to 5200 Hz or 5300 Hz, depending on the type of IC2 used (as indicated in the service manual).

Pilot deviation sensitivity adjustment. Set the FM generator RF output for 2 mV. Externally modulate the FM generator with 5300 Hz. Adjust the generator deviation setting for ±1 kHz. Connect the oscilloscope d-c input to TP6 (Fig. 5-6). If the oscilloscope shows approximately 8 V dc, slowly adjust VR2 to the point at which the oscilloscope d-c voltage level changes to zero.

5-5.4 Portable-Unit Receiver Alignment

Preparation for alignment. Disconnect speaker SP401 (Fig. 5-16) and connect an 8-Ω load resistor across the output. Disconnect the telescoping antenna. Connect all of the test equipment (including the resistive network) to the portable unit as shown in Fig. 5-29. Set the HIGH/LOW switch S304 to HIGH, the power ON/OFF switch S302 to ON, and the TALK/STANDBY switch S301 to TALK.

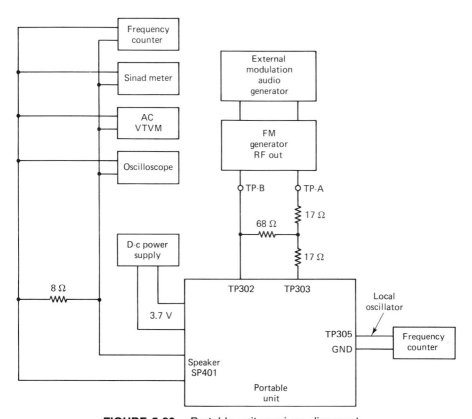

FIGURE 5-29. Portable-unit receiver alignment

Local oscillator frequency adjustment. Connect the frequency counter to TP305 (Fig. 5-16). Measure the crystal oscillator frequency. If the frequency exceeds ±100 Hz from the specified crystal X301 frequency, adjust T306 for the correct frequency. (Check the service literature for the correct X301 frequency.)

Sensitivity adjustment. With the FM generator connected as shown in Fig. 5-29, set the modulating frequency to 1 kHz and the deviation to ±2 kHz. Adjust the FM generator RF output until a reading of 12 dB is obtained on the Sinad meter. This should occur with an output of 2 μV or less for a properly aligned portable-unit receiver. Skip the alignment if the Sinad reading is 12 dB with an output from the FM generator of 2 μV or less. (Why look for trouble?)

If alignment is required, first adjust T307 for maximum audio on the AC VTVM. Then adjust T302, T303, and T304 for minimum reading on the Sinad meter while maintaining maximum audio output on the AC VTVM. During alignment, keep reducing the generator output to maintain a 12-dB indication on the Sinad meter.

Connect the high-impedance RF voltmeter to TP-A and TP-B on the resistive network (Fig. 5-29). Carefully adjust T312 (Fig. 5-16) for the lowest reading on the Sinad meter while maintaining optimum RF power output on the RF meter (so that the Sinad reading is minimum, but near 12 dB).

5-5.5 Portable-Unit Transmitter Alignment

Preparation for alignment. Disconnect the telescoping antenna. Connect all of the test equipment (including the resistive network) to the portable unit as shown in Fig. 5-30. Set the power ON/OFF switch S302 to ON and the TALK/STANDBY switch S301 to TALK.

Transmitter power output adjustment. Measure the RF voltage across TP-A and TP-B of the resistive network. The RF voltmeter should read between 150 and 250 mV. Adjust T310, T311, T309, T312, and T308 (Fig. 5-18), in that order, for maximum reading on the RF voltmeter.

Repeat the alignment of T310, T308, T309, and T312, in that order. Then align T311 for maximum power. Retest the receiver for 12-dB Sinad sensitivity, as described in Sec. 5-5.4. If the sensitivity has degraded, carefully adjust T312 for best signal/noise and power output.

Transmitter frequency and pilot deviation. Disable the 5.3-kHz pilot tone generator by shorting TP307 and TP306 (Fig. 5-23). Measure the transmitter frequency with the frequency counter. The transmitter frequency should not exceed ±100 Hz from the designated channel frequency (Fig. 2-9). Adjust L303 and TC302 (Fig. 5-18) if the transmitter frequency is not within tolerance.

Remove the short between TP306 and TP307 to turn on the 5.3-kHz pilot-tone generator. The transmitter frequency should now show a deviation (caused by the pilot tone) between ±2.3 and ±3.7 kHz. If necessary, adjust TC302 and L303 until the correct frequency and deviation are achieved.

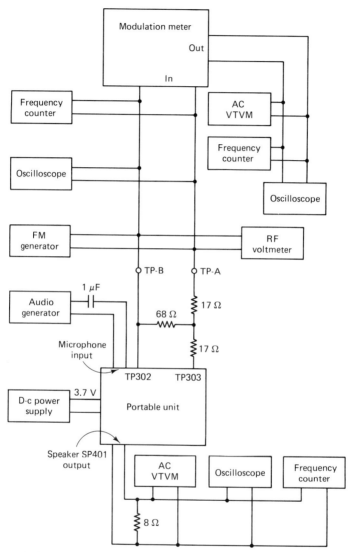

FIGURE 5-30. Portable-unit transmitter alignment

Repeat the adjustment for maximum power output by adjusting T308, T309, T310, and T311. Then correct the transmitter frequency by readjusting L303, as necessary.

Pilot frequency adjustment. If the transmitter does not produce the correct deviation, the problem may be with the 5.3-kHz pilot tone generator. Connect the frequency counter to the output of the modulation meter and measure the frequency. If necessary, adjust VR304 (Fig. 5-23) to produce the correct 5.3-kHz deviation.

Modulation sensitivity check. Disable the 5.3-kHz pilot tone generator by connecting TP307 to TP306 (Fig. 5-23). Disconnect the microphone (Fig. 5-18). Connect an audio generator in place of the microphone, with a 1-µF electrolytic capacitor in series with the microphone lead (negative terminal of the capacitor to the generator). Connect the modulation meter to TP-A and TP-B of the resistive network.

Set the audio generator frequency to 1 kHz. Increase the audio generator output until the modulation meter indicates ±2 kHz deviation. The audio generator output should be between 12 and 18 mV. Connect the oscilloscope to the modulation meter output and observe the 1-kHz signal for nonlinearity or distortion. Make certain to remove the capacitor before reconnecting the microphone when the check is complete.

Hook-switch flash time adjustment. Connect the oscilloscope to the base-unit relay RL1 (Fig. 5-9). Turn the portable unit on and capture the telephone line (the base-unit IN-USE indicator LED should be on). Press the FLASH button (momentarily) on the portable-unit keypad (Fig. 5-26). Measure the time interval that the base-unit relay RL1 remains off. The time should be between 400 and 800 ms. If necessary, adjust VR301 to obtain the correct flash interval.

Digital code signal deviation check. Disable the 5.3-kHz pilot-tone generator by connecting TP307 to TP306 (Fig. 5-23). Connect the modulation meter to TP-A and TP-B of the resistive network. Connect pin 14 of IC402 (Fig. 5-23) to ground (so that the digitally coded signal at IC402-15 remains on). The modulation meter should read between ±3 and ±5 kHz deviation.

To check the actual code generator output, connect the oscilloscope to pin 15 of IC402. The digital output should be approximately as shown in Fig. 5-13 (8 V peak).

2.5-kHz call-signal oscillator check. Connect the collector of Q314 (Fig. 5-21) to ground (to turn on the oscillator). Connect the oscilloscope to the oscillator output (at the junction of IC403-6 and R394) and monitor the 2.5-kHz signal.

5-5.6 Portable-Unit Low-Battery Indicator Check

Figure 5-31 shows the portable-unit battery circuits, including the low-battery indicator. The BATT LOW indicator LED D454 turns on when the charge on the portable-unit nickel-cadmium battery goes below a certain level (between 3.2 and 3.5 V).

When the portable unit is on the base-unit cradle, the 3.6-V battery is charged from the power supply circuits in the base unit, through terminals J413 and J414. The battery can also be charged through EXT DC connector J406 by an external or auxillary d-c source. The charging current is removed when the portable unit is lifted from the base unit.

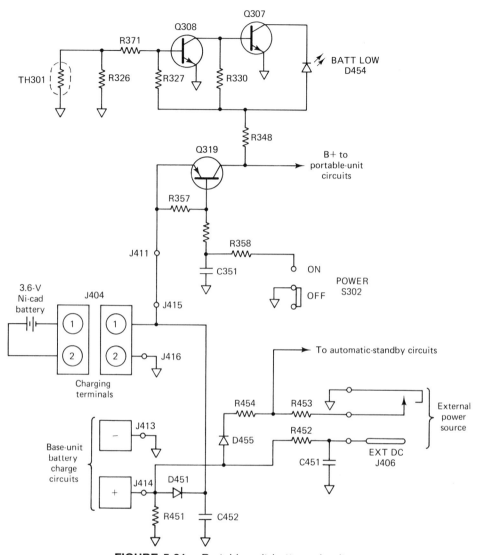

FIGURE 5-31. Portable-unit battery circuits

The battery power is distributed to all of the portable-unit circuits through J411 and Q319. Transistors Q307 and Q308, which drive D454, also receive power from J411.

When the power-circuit (or B+) voltage is 3.6 V or higher (battery fully charged), Q307 is cut off, keeping the cathode of D454 at a high impedance (D454 is reverse biased and off). When the power-circuit voltage drops to a level between 3.2 and 3.5 V (battery charge low), Q307 turns on, connecting the cathode of D454 to ground. Since the anode of D454 is connected to the power-circuit voltage line, D454 turns on to indicate a low battery charge.

To check operation of the low-battery circuits, remove the battery and connect an external d-c power supply to J415(+) and J416 (−). (Set the power supply to zero *before* connecting to the battery-circuit terminals!)

Slowly increase the external power supply from zero to 3.6 V. The BATT LOW indicator D454 should turn on, and then off, when the voltage reaches 3.6 V.

Slowly decrease the power supply from 3.6 V. The BATT LOW indicator D454 should turn on at some voltage between 3.2 and 3.5 V. Make certain to disconnect the external source before replacing the battery.

5-6. INTRODUCTION TO CORDLESS TELEPHONE TROUBLESHOOTING

Figure 5-32 is a trouble symptom chart for the base unit of a cordless telephone. Figure 5-33 is a similar trouble chart for a cordless telephone portable unit. The telephones selected are those described in Secs. 5-2 through 5-4.

The charts provide lists of various symptoms common to most cordless telephones. After selecting the symptom which matches that of the cordless telephone being serviced, follow the steps in the corresponding troubleshooting procedures (indicated by section number in the right-hand column of Figs. 5-32 and 5-33). These procedures help isolate the problem to a defective circuit or component.

Listed in each troubleshooting procedure are the adjustments associated with the related circuit. The adjustments are described in Sec. 5-5. Always check the adjustments, if possible, before proceeding with service or troubleshooting. You might just find the trouble!

Some of the components described in the troubleshooting sections are given in the diagrams of Figs. 5-3 through 5-26, and 5-31. The troubleshooting sections refer to the illustrations that show the components (as well as the adjustment and measurement points). Additional schematic diagrams are referenced from the troubleshooting sections as required to understand the troubleshooting procedures.

Symptom	Troubleshooting Procedure Section
Base unit dead, POWER indicator does not turn on	5-7
Telephone line cannot be seized	5-8
Telephone does not ring	5-9
Voice from portable unit cannot be heard on telephone line	5-10
Portable unit rings but calling party cannot be heard	5-11
Base unit does not pass through dial signals	5-12
No paging buzzer (call signal) when CALL button is pressed	5-13
IN-USE indicator does not turn on	5-14
CHARGE indicator does not turn on	5-15

FIGURE 5-32. Base-unit trouble symptom chart

Symptom	Troubleshooting Procedure Section
Portable unit dead	5-16
Battery does not charge with portable unit in cradle	5-17
Battery does not charge from external source	5-18
BATT LOW indicator does not turn on and off properly	5-19
TALK indicator does not turn on	5-20
Portable unit does not ring	5-21
Telephone line cannot be seized or base unit activated	5-22
No dialing or redial function	5-23
Telephone rings but no sound on portable-unit speaker	5-24
Telephone rings but voice from portable unit cannot be heard	5-25
Flash (call-waiting) not operative	5-26
Automatic standby function not operative	5-27
No keystroke pulse generated when dial keys are pressed	5-28

FIGURE 5-33. Portable-unit trouble symptom chart

5-7. BASE UNIT DEAD, POWER INDICATOR DOES NOT TURN ON

Figure 5-34 shows the circuits involved. There are no adjustments.

If the base unit appears to be dead and the POWER indicator D203 does not turn on, the first step is to check fuse F1 (right after you make sure that the power cord is plugged in!). Then check the voltage at the secondary of T13.

Next check the output of D15 on both sides of R122. Typically, the collector of Q26 is 13.3 V. The base of Q26 is 8.74 V, as set by zener D16.

The output from the series regulator at the emitter of Q26 is 9.39 V. Note that this output is the main B+ power source for the base-unit circuits.

If the B+ at the collector of Q26 appears to be good but the POWER indicator is not on, the most likely suspects are D203, R203, or possibly a poor connection at pin 8 of J2 (or simply broken B+ wiring).

5-8. TELEPHONE LINE CANNOT BE SEIZED

If an assumed-good portable unit cannot seize the base unit (IN-USE indicator does not turn on, telephone line is not connected, etc.), the problem is most likely in the base unit (although the problem can be in the portable unit; refer to Sec. 5-22).

Start by checking that the POWER indicator is on and that the base unit is connected to a telephone output (this always helps). Next, check that the positions of code switch S202 match the code-switch positions on the portable unit. If all of the obvious problems are eliminated but the line cannot be seized, you must first

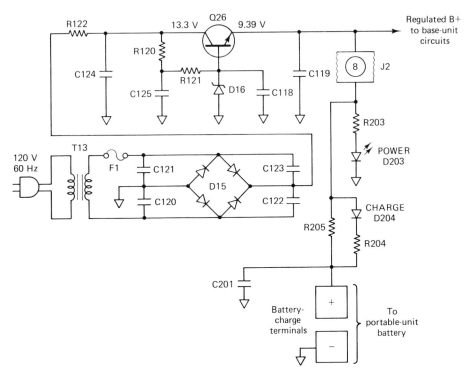

FIGURE 5-34. Base-unit power circuit troubleshooting diagram

determine if the problem is in the base unit or in the portable unit. Substitution is the easiest solution, but not always practical.

If a known-good portable unit cannot seize the line and base unit, there are two basic possibilities. First, the base-unit receiver may not be passing the 5.3-kHz pilot tone and/or the security-code signal from the portable unit. The second possibility is that both signals are available at the receiver output but are not being processed properly.

Figure 5-35 shows the circuits involved in processing the pilot and security signals. Figure 5-36 shows the circuits used to turn on the base unit.

Start by checking pins 3 and 11 of IC4, shown on both Figs. 5-35 and 5-36. IC4-3 should be high, with IC4-11 low (the IC4 FF properly set), for the base unit to be turned on and the telephone line seized.

If the IC4 FF is properly set but the base unit is not seized, check the following (Fig. 5-36):

Check that there is B+ at the collectors of Q7 and Q12. If not, suspect Q20.

Check that the IN-USE indicator D201 is on. If not, suspect Q19 and IC5.

Check that the telephone line is connected to the hybrid transformer. If not, suspect RL1, Q21, and IC6. Note that TP7 should be high if D201 is on. If not, suspect IC6.

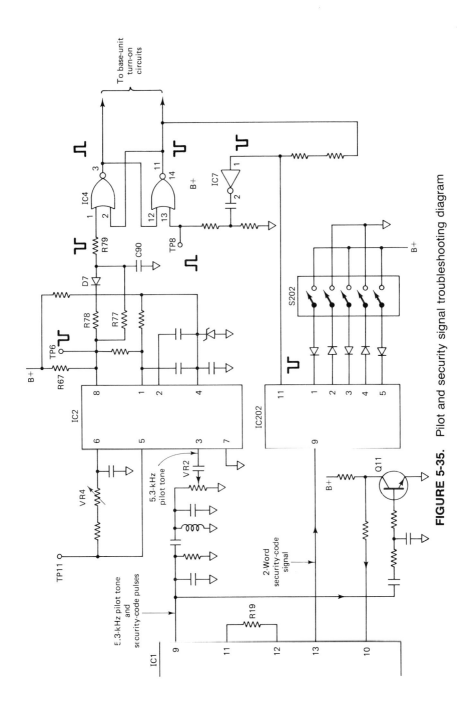

FIGURE 5-35. Pilot and security signal troubleshooting diagram

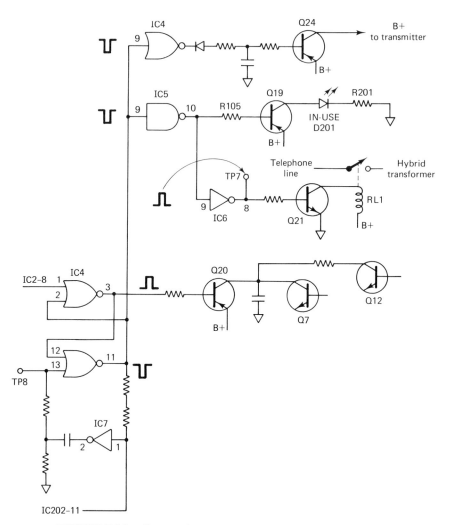

FIGURE 5-36. Base-unit turn-on-circuit troubleshooting diagram

If the IC4 FF is not properly set, check the following (Fig. 5-35):

Check at IC1-9 (with an oscilloscope) for a 5.3-kHz signal and for security-code pulses. You probably cannot decipher the pulse code. However, the presence of both signals indicates that the receiver circuits are probably good. If the signals are absent or you suspect the receiver circuits, try running through the receiver alignment procedures described in Sec. 5-5.3. This will show up any basic problems in the receiver. Make the usual voltage and/or resistance checks at the terminals of the receiver transistors.

Next, check for a low (zero volts) at pin 8 of IC2 (or at TP6). A low indicates that the 5.3-kHz signal is present at IC2-3 (and is of correct amplitude) and that the frequency matches that set by VR4.

If IC2-8 is not low, make sure that there is a 5.3-kHz signal at IC2-3 and that there is a signal of matching frequency at TP11. If the signal is absent or abnormal at IC2-3, check the components between IC2-3 and IC1-9 as well as the adjustment of VR2 (Sec. 5-5.3). If the signal at TP11 is absent or abnormal, check the associated components and the adjustment of VR4 (Sec. 5-5.3).

If IC2-8 is not low but the 5.3-kHz signals are good, suspect IC2 or the associated component.

If there is a low at IC2-8, check for a low at IC4-1. If IC4-1 is not low, check R77, R78, R79, D7, and C90.

Next, check for a low at pin 11 of IC202 and a high at IC4-13 (or at TP8). The low at IC202-11 indicates that the security-code signal is present at IC202-9 and that the portable-unit code matches the base-unit code selected at S202.

If IC202-11 is not low, check for the presence of pulses (the security-code signal) at pins 9, 10, 11, 12, and 13 of IC1.

If the pulses are present at pin 9 but not at pin 10, suspect Q11.

If the pulses are present at pin 10 but not at pin 11, suspect IC1.

If the pulses are present at pin 11 but not at pin 12, suspect R19.

If the pulses are present at pin 12 but not at pin 13, suspect IC1.

If the pulses are present at IC202-9 but IC202-11 is not low, IC202 is suspect. However, S202 or the components between IC202 and S202 could be the problem. (Before you tear into the components, make absolutely sure that S202 is set to the *same code* as the portable unit!).

If IC202-11 is low but IC4-13 (TP8) is not high, suspect IC7.

With pin 1 of IC4 low and IC4-13 high, pin 3 of IC4 should latch high while IC4-11 should go low. *Both of these conditions* must occur to seize the base-unit circuits shown in Fig. 5-36.

5-9. TELEPHONE DOES NOT RING

If the telephone does not ring when the proper number is dialed from another telephone or when a test ring signal is applied to the base-unit telephone input (with an assumed-good portable unit; Sec. 5-21), there are two basic possibilities.

First, the base-unit transmitter may not be modulated by the 2.1-kHz ring/call signal or the security-code signal. Second, both signals may be available to the base-unit transmitter, but the transmitter may not pass the signals to the portable unit.

If you suspect the transmitter circuits, try running through the transmitter alignment procedure described in Sec. 5-5.2. This will show up any basic problems in the transmitter. Make the usual voltage and/or resistance checks at the terminals of the transmitter transistors.

If the transmitter is good, check the ring circuits. Figure 5-37 shows the circuits involved.

The first point to check is the neon coupler OC1. Pin 6 of IC4 and pin 13 of IC5 should go low when a ring signal is present on the telephone line. If not, suspect OC1.

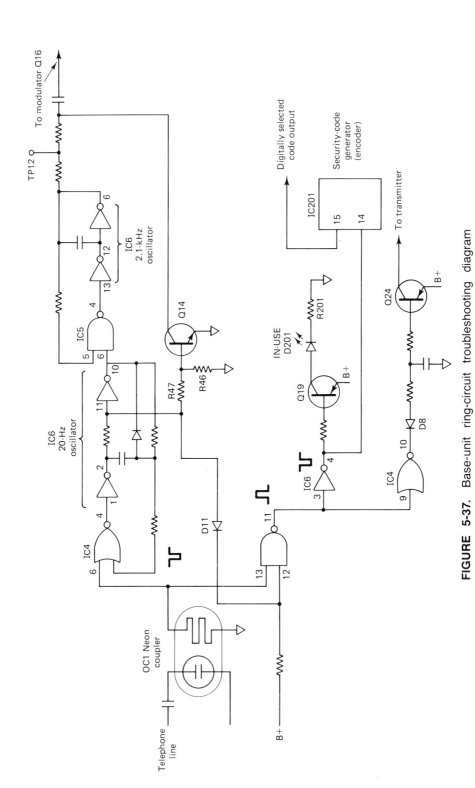

FIGURE 5-37. Base-unit ring-circuit troubleshooting diagram

If OC1 is good, check for a 2.1-kHz signal at TP12 whenever the ring signal is applied to the telephone line. If the signal is absent, suspect D11, Q14, IC4, IC5, and IC6. Q14 should short the 2.1-kHz output, except when the ring signal is present.

If the 2.1-kHz signal is present and applied to the modulator, check that IC5-11 goes high when the ring signal is applied. If not, suspect IC5.

With IC5-11 high, IC6-4 should go low, turning on Q19 and IN-USE indicator D201. If not, suspect IC6, Q19, or D201.

With IC5-11 high, IC4-10 and the cathode of D8 should go low, turning on Q24 and the transmitter. If not, suspect IC4, D8, or Q24.

With IC6-4 (and IC201-14) low, Q201 should turn on, producing pulses (security code) at IC201-15. If not, suspect IC201. Again, you need not decipher the code, but the presence of pulses indicates that IC201 is probably good.

5-10. VOICE FROM PORTABLE UNIT CANNOT BE HEARD ON TELEPHONE LINE

The procedures described thus far in this chapter establish that the telephone is not dead (Sec. 5-7), that the telephone line and base unit can be seized by the portable unit (Sec. 5-8), and that the telephone rings when called or a test ring signal is applied (Sec. 5-9). These procedures prove that the transmitter and receiver of both the portable and base units are probably good. So if voice cannot be heard on another telephone (with an assumed-good portable unit; Sec. 5-25), the problem is likely to be in the base-unit voice circuits.

Figure 5-38 shows the circuits involved. As in the case of any audio circuits, the simplest way to locate problems is to trace the audio signals through the circuits with an oscilloscope (or other audio signal tracer).

With the telephone turned on (TALK button pressed), speak into the portable-unit microphone (or inject an audio signal at the microphone input) and trace the audio from pin 9 of IC1 through to transformer T8. Keep the following in mind when tracing audio through the circuits of Fig. 5-38.

Only voice audio should appear at points from the output of VR1 through Q8/Q9 and T8. Voice audio, security-code pulses, and the 5.3-kHz pilot signal appear at IC1-9.

If there is no audio at the emitter of Q7, check that Q7 is turned on through Q20 by a valid transmission signal (low) from pin 11 of IC202. If the IC202-11 signal is not low, suspect IC202 (or improper setting of S202, which you should have checked first!).

Also, the low signal that turns Q20 (and Q7) on turns mute transistor Q27 off, permitting audio to pass. If Q27 is on (when there is no valid transmission when the settings of S202 do not match those of the portable unit), the audio is shorted to ground. Of course, if the base-unit transmitter is on (IN-USE indicator on, etc.), it is reasonable to assume that there is a valid transmission signal at

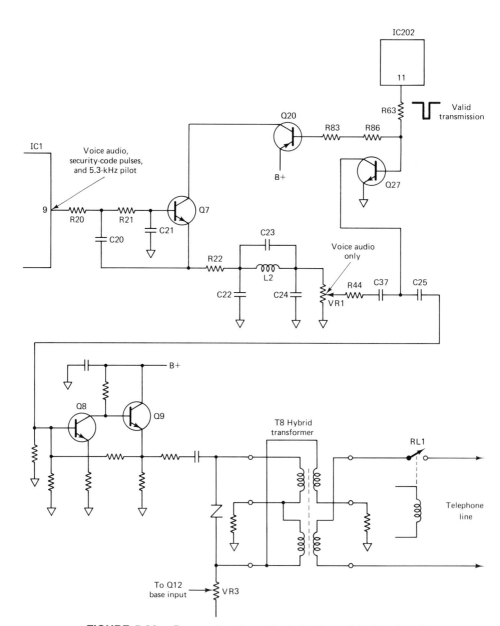

FIGURE 5-38. Base-unit voice-output-circuit troubleshooting diagram.

IC202-11. So, if Q27 is on, look for a short from the base of Q27 to B+ (or a similar condition).

5-11. PORTABLE UNIT RINGS BUT CALLING PARTY CANNOT BE HEARD

If the telephone rings when dialed by another telephone but voice cannot be heard on the portable unit (with an assumed-good portable unit; Sec. 5-24), the problem is likely to be in the voice input circuits.

Figure 5-39 shows the circuits involved. Start by checking the adjustment of VR3 as described in Sec. 5-5.2. Next, check Q12 and the associated components. Keep in mind that Q12 receives B+ though Q20, which must be turned on by a low at IC202-11.

Also check transformer T8. It is possible that some of the windings are defective even though other good windings will pass a ring signal from the telephone line input.

Keep in mind that this same symptom can be caused by problems in the portable unit (refer to Sec. 5-24).

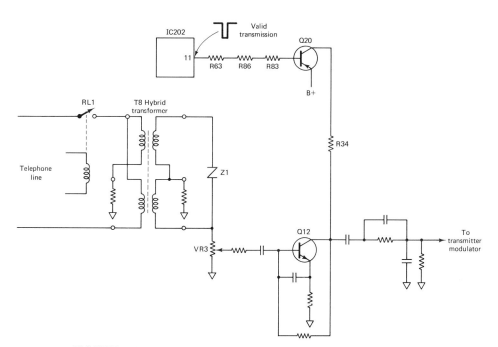

FIGURE 5-39. Base-unit voice-input-circuit troubleshooting diagram

5-12. BASE UNIT DOES NOT PASS THROUGH DIAL SIGNALS

If a cordless telephone fails to dial, it is most likely to be the fault of the portable unit (refer to Sec. 5-23.) However, it is possible that the base-unit circuits do not operate properly when pulse-dialed by the portable unit (as described in Sec. 5-4.3) even though the same circuits are good when not being dialed.

The circuits involved are shown in Figs. 5-6 and 5-9. If there are no pulses at IC1-9 when the portable unit is pulse dialed, suspect the receiver circuits. Try the alignment procedures of Sec. 5-5.3.

If there are dialing pulses at IC1-9, trace these pulses from IC2-3 through to relay RL1, as illustrated in Fig. 5-9.

5-13. NO PAGING BUZZER (CALL SIGNAL) WHEN CALL BUTTON IS PRESSED

Figure 5-13 shows the circuits involved. Keep in mind that both the 2.1-kHz signal and the security-code pulses must be applied to the modulator, and the transmitter must be turned on, before the call signal is transmitted to the portable unit. All of these conditions should occur when CALL switch S201 is pressed.

If the 2.1-kHz signal is not present at C72 when S201 is pressed, look for the 2.1-kHz signal at TP12. If missing, suspect IC5, IC6, C72, C80, R89, R90, R97, and R110.

Next, press S201 and check for a low (or ground) at the cathode of D11. If not, suspect S201.

Check that the base of Q14 goes low when S201 is pressed. This should also cause Q14 to turn off, so that the collector Q14 rises to about 2.5 V and allows the 2.1-kHz signal to pass. If not, suspect R46, R47, and Q14.

If the security-code pulses are not present at C46, press S201 and check for pulses at TP13. If present at TP13, but not at C46, suspect IC5 or R98.

If pulses are absent at TP13, press S201 and check for a low at IC5-12, a high at IC5-11, IC5-2, and IC6-3, and a low at IC6-4 and IC201-14.

IC201-14 must go low for pulses to be produced at IC201-15. If IC201-14 is not low, with S201 pressed, suspect IC6 and IC5.

Finally, check that the collector of Q24 goes to the same voltage as the base (about 8.75 V) when S201 is pressed (to produce a high at IC4-9 and a low at IC4-10). If not, suspect IC4, D8, R101, R102, C82, and Q24.

5-14. IN-USE INDICATOR DOES NOT TURN ON

Figures 5-6 and 5-12 show the circuits involved. If the IN-USE indicator D201 does not turn on with power indicator D203 on, check that there is a low at the base of Q19 and B+ present at the collector of Q19. If the base of Q19 is not low, check for a high at IC5-9 and IC4-3, indicating that the IC4 flip-flop is set to turn the transmitter on.

Of course, if the base of Q19 is low and the collector of Q19 is at B+ but D201 does not turn on, suspect D201 or R201.

5-15. CHARGE INDICATOR DOES NOT TURN ON

Figure 5-34 shows the circuits involved. CHARGE indicator D204 should turn on when there is sufficient current passing through R205 (indicating that the portable-unit battery is charging). If not, first check that POWER indicator D203 is on and that the portable unit is properly seated in the cradle. The battery-charging contacts on the portable unit must mate properly with the base-unit contacts.

If the portable unit is in place and the POWER indicator D203 is on but D204 does not turn on, suspect D204 or R204. It is also possible that the portable-unit battery is defective (which we discuss next in Sec. 5-16).

5-16. PORTABLE UNIT DEAD

If the portable unit appears to be completely dead but the base unit is good (portable-unit OFF/ON switch set to ON; base-unit POWER indicator on), try substitution if practical. If substitution is not practical, check the portable-unit battery (again by substitution). Keep in mind that the battery may be good but not properly charged. Check the BATT LOW indicator. If you suspect the battery, refer to Secs. 5-17 through 5-19. If the portable unit is dead with a known-good and properly charged battery, check the power-control circuits.

Figure 5-40 shows the circuits involved.

When POWER switch S302 is set from OFF to ON, the base of Q319 is

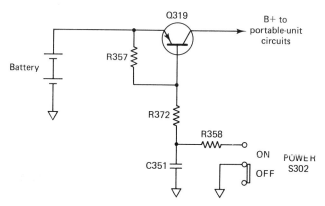

FIGURE 5-40. Portable-unit battery/B+ circuit troubleshooting diagram

returned to ground through R358 and R372. This turns Q319 on and connects the battery plus (+) terminal to the B+ circuits within the portable unit.

If there is no B+ at the collector of Q319 with S302 set to ON but there is B+ at the emitter of Q319, suspect S302, Q319, R358, and R372. Also check C351 (for possible shorts or leakage) and R357.

5-17. BATTERY DOES NOT CHARGE WITH PORTABLE UNIT IN CRADLE

Figure 5-41 shows the circuits involved. First, check the battery for a voltage of about 3.6 V. If the battery voltage is very low, the battery may be defective. Try a new battery.

If a known-good battery does not charge, suspect D451, or possibly R451 and C452. (When the battery voltage drops to a certain value, D451 is forward biased.)

5-18. BATTERY DOES NOT CHARGE FROM EXTERNAL SOURCE

Figure 5-42 shows the circuits involved. First, make sure that the battery charges from the base-unit terminals (portable unit in cradle) as described in Sec. 5-17. If so but the battery does not charge from an external source, the problem is likely to be in the contacts of EXT DC connector J406. Also check R452, R453, R454, C451, C452, and D455. Note that a defect in some of these components can also affect the automatic-standby function, described in Sec. 5-27.

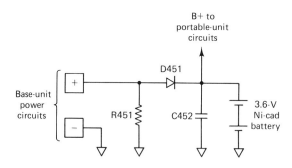

FIGURE 5-41. Portable-unit battery-charge-circuit troubleshooting diagram

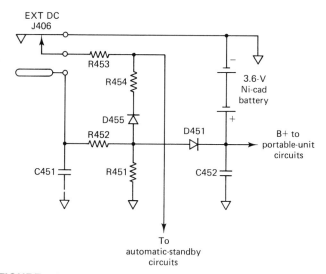

FIGURE 5-42. Portable-unit external-charge-circuit troubleshooting diagram

5-19. BATT LOW INDICATOR DOES NOT TURN ON AND OFF PROPERLY

Figure 5-43 shows the circuits involved. If BATT LOW indicator D454 turns on with a known-good and fully charged battery, or does not turn on with a low battery, suspect D454, Q307, and Q308.

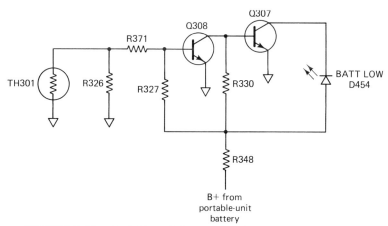

FIGURE 5-43. Portable-unit battery-low (BATT LOW)-circuit troubleshooting diagram

When the battery voltage or B+ is normal, Q308 is turned on and Q307 is turned off (by the drop across R330). When battery or B+ is low, Q308 turns off, reducing the drop across R330 and turning Q307 on. This returns the cathode of D454 to ground (through Q307 and R325) and forward biases D454.

5-20. TALK INDICATOR DOES NOT TURN ON

The TALK indicator D453 is controlled by the standby/talk circuits, described in Sec. 5-22.

5-21. PORTABLE UNIT DOES NOT RING

If the telephone does not ring when the proper number is dialed from another telephone, or when a test ring signal is applied to the base-unit telephone input (with an assumed-good base unit; Sec. 5-9), there are two basic possibilities.

First, the portable-unit receiver may not be passing the 2.1-kHz ring/call signal or the security-code signal. Second, both signals may be available at the output of the portable-unit receiver, but the signals are not being processed properly.

If you suspect the receiver circuits, try running through the receiver alignment procedure described in Sec. 5-5.4. This shows up any basic problems in the portable-unit receiver. Make the usual voltage and/or resistance checks at the terminals of the receiver transistors.

If the receiver is good, check the ring circuits. Figure 5-21 shows the circuits involved.

First, check IC401-11 for a high when a ring or call signal is applied. If IC401-11 is low, check for pulses at IC401-9. If the pulses are present and S401 is set to the proper code (which you should have checked in the first place), suspect IC401.

If the security code pulses are absent at IC401-9, check for pulses at IC301-11. If present at IC301-11 but absent at IC401-9, suspect Q324, C325, R329, and R396.

Note that Q324 is controlled by STANDBY/TALK switch S301. The circuit involved are discussed in Sec. 5-22.

If the pulses are absent at IC301-11, the 2.1-kHz ring/call signal is probably absent at IC301-9, and the portable-unit receiver circuits are suspect.

If Q325 is turned on (by the high at IC401-11), trace the 2.1-kHz ring/call signal from IC301-9 through to Q314. The signal should be amplified by Q325, and rectified by Q326, to turn Q314 on.

Next, check that there is a 2.5-kHz call or paging signal at IC403-6, at the base of Q322. If the signal is absent at IC403-6, suspect IC403, C415, R415, R417, and D410. If the signal is present at IC403-6 but absent at Q322, suspect

R394. If the signal is present at the base of Q322, but absent at SP402, suspect Q322, R393, and SP402.

Note that if IC401-11 is low (no valid transmission), D305 is forward biased, and the 2.5-kHz paging signal does not pass to Q322.

5-22. TELEPHONE LINE CANNOT BE SEIZED OR BASE UNIT ACTIVATED

If an assumed-good base unit cannot be seized, the problem is likely to be in the portable unit (although the problem can be in the base unit; refer to Sec. 5-8).

Start by checking that the OFF/ON switch is ON and that the BATT LOW indicator is not on. Next, check that the positions of the portable-unit code switch S401 match the code switch positions of the base-unit switch S202.

If all of the obvious problems are eliminated but the line cannot be seized or the base unit activated, you must first determine if the problem is in the base unit or in the portable unit. Substitution is the easiest solution but is not always practical.

If a known-good base unit cannot be seized, there are two possibilities. First, the portable-unit transmitter may not be passing the 5.3-kHz pilot-tone and/or security-code signal. The second possibility is that either or both signals are not available at the portable-unit transmitter input.

If you suspect the transmitter circuits, try running through the transmitter alignment procedures described in Sec. 5-5.5. This shows up any basic problems in the portable-unit transmitter. Make the usual voltage and/or resistance checks of the terminals of the transmitter transistors.

Keep in mind that the portable unit must be in the *standby condition* (STANDBY/TALK switch S301 in STANDBY) to receive a call. For example, if there is a ring signal applied to the base unit, or the CALL button is pressed, the base-unit transmitter is modulated (for the duration of the ring, or while the CALL button is pressed) by the 2.1-kHz ring/call signal and a security-code signal. These signals are picked up by the portable-unit receiver and produce a 2.5-kHz paging or call signal on ring speaker SP402.

When S301 is set to TALK, the portable-unit transmitter is turned on and modulated by a 5.3-kHz pilot signal as well as the security-code signal. In turn, these signals are transmitted to the base-unit receiver and cause the base-unit flip-flop to latch (or set) so that the base-unit transmitter stays on. This completes the communciations link between the base and portable units.

Figure 5-44 shows the standby/talk circuits. When the STANDBY/TALK switch S301 is in STANDBY and POWER switch S302 is ON, the cathode of D303 is grounded. This connects the base of Q317 to ground through R353 and R363. Q317 turns on and passes battery B+ (from Q319) to the transmitter circuits. (Q319 turns on to pass battery B+ when S302 is set to ON.)

The ground at the cathode of D303 also connects the base of Q318 to

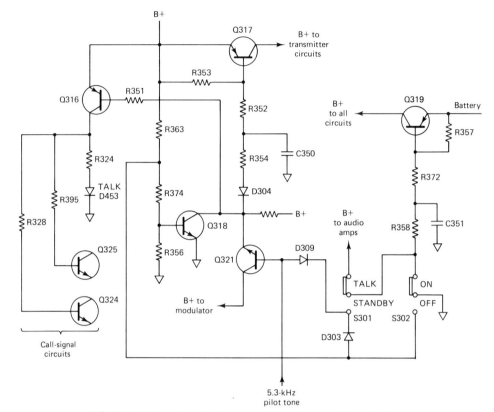

FIGURE 5-44. Standby/talk-circuit troubleshooting diagram

ground through R374. This turns Q318 off, causing the collector of Q318 to rise and turn off Q316. With Q316 off, the call-signal circuits (Q324, Q325, and Q326; Fig. 5-21) are placed in a condition to operate (as described in Sec. 5-21).

When S301 is moved from STANDBY to TALK and S302 is ON, the cathode of D303 is removed from ground. This removes the ground from the base of Q318, causing Q318 to turn on (because of the B+ through R363). With Q318 on, the collector voltage drops (to almost zero). This low connects the cathode D304 to ground. Q317 remains on and passes battery B+ (from Q319) to the transmitter circuits.

With Q318 on (collector low), Q316 turns on, turning off the call circuits (Fig. 5-21). Also, the emitter of Q321 is returned to ground through Q318, placing Q321 in a condition to pass the 5.3-kHz pilot tone signals to the transmitter modulator circuits (Fig. 5-23).

Moving S301 to TALK also removes the ground from the cathode of D309, permitting the 5.3-kHz signals to pass.

If the portable unit rings when the base-unit CALL button is pressed, it can be assumed that the portable-unit receiver and call-signal circuits (Fig. 5-21) are good. If not, follow the procedures of Sec. 5-21.

Once you have determined that the portable-unit ring circuits are good, the first step in troubleshooting the communications-link circuits is to set S301 to TALK and check that the TALK indicator D453 turns on. If so, Q316, Q317, and Q318 are good. If not, check Q316, Q317, and Q318. All three transistors should be turned on when S302 is in TALK and S301 is ON. Q318 is turned on by B+ through R363 and R374. The collector of Q318 should drop to almost zero, turning on Q316 (through R351) and Q317 (through D304, R354, and R352).

If TALK indicator D453 turns on, next check that there is a digital signal (pulses) at IC402-15 and a 5.3-kHz signal at the base of Q321, as shown in Fig. 5-45.

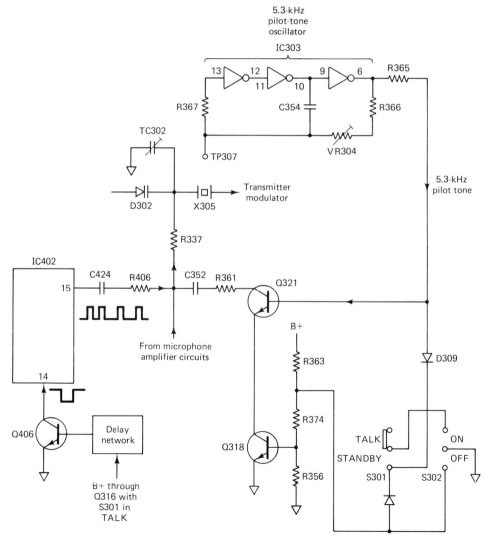

FIGURE 5-45. Modulator-input-circuit troubleshooting diagram

172   Cordless Telephone Troubleshooting

If there are no pulses at IC402-15 (after a slight delay), suspect Q406 or the delay network components. Note that the delay network and Q406 receive B+ through Q316 when S301 is in TALK, but IC402-14 does not go low immediately.

If there are pulses at IC402-15, check that the pulses are applied to varicap D302 in the transmitter modulator (through C424, R406, and R337).

If there is no 5.3-kHz signal at the base of Q321, check the pilot-tone generator or oscillator components IC303, VR304, R365, R366, R367, and C354.

If necessary, adjust VR304 as described in Sec. 5-5.5. Also check that the ground is removed from the cathode of D309 when S301 is in TALK.

If there is a 5.3-kHz signal at the base of Q321, check that the signal is applied to varicap D302 (through R361, C352, and R337). Note that the emitter of Q321 is returned to ground through Q318 *only* when Q318 is turned on (by S301 being in TALK).

5-23. NO DIALING OR REDIAL FUNCTIONS

Before troubleshooting the dial or redial functions, make certain that the portable unit rings (Secs. 5-9 and 5-21) and that the telephone line and base unit can be seized (Secs. 5-12 and 5-22).

Figure 5-46 shows the circuits involved for a pulse-dial telephone. Monitor IC405-18 for dial pulses when any of the keys on keypad S405 are pressed. If there are no pulses at IC405-18, or an incorrect number of pulses, suspect IC405 or keypad S405 or the wiring between the two components.

Next, check the IC405 external components, such as C408, C412, R415, and R416. Note that IC405-17 must be low for pulses to be produced at IC405-18. The low is provided when Q401 is turned on by B+ through Q316 (when S301 is in TALK; Sec. 5-22). During standby, Q401 is off and IC405-17 is held high by the B+ through R410.

If there are pulses at IC405-18, check that the pulses are applied to the base of Q321 through R364, Q309, and R362. The pulses should turn Q321 on and off, interrupting the 5.3-kHz pilot tone. If the pulses do not reach Q321, suspect Q309 and the associated circuits.

Figure 5-47 shows the circuits involved for combination pulse-dial and tone-dial telephones. The tones are produced at IC404-16 under control of crystal X401 and are applied to the varicap D302 in the transmitter modulator circuits through C411 and R333. The tones are also applied to the speaker through the audio circuits (Sec. 5-24).

IC404-2 must be high to turn IC404 on. When PULSE/TONE switch S402 is in PULSE, IC404-2 is low (due to the ground at the PULSE terminal of IC404-2). This turns IC404 off and prevents tones from being produced at IC404-16. IC405 then produces pulses as shown in Fig. 5-46. (IC405-17 is held low through the PULSE terminal of S402-1 and Q401.)

No Dialing or Redial Functions 173

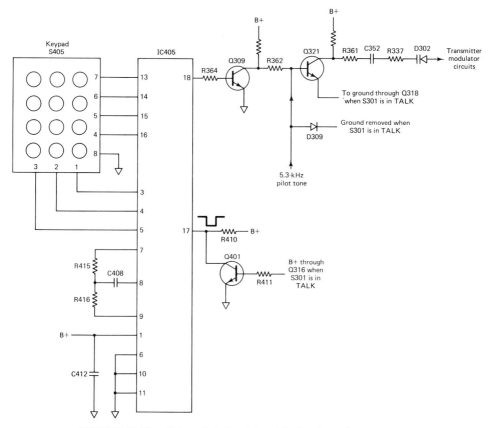

FIGURE 5-46. Pulse-dial-circuit troubleshooting diagram

When S402 is in TONE, Q401 remains turned on (by B+ through Q316 in TALK), and the collector of Q401 remains low, turning Q407 off. This removes the low at IC404-2 and turns IC404 on to produce tones at IC404-16. The correct tone pairs are determined by which buttons are pressed on keypad S405.

With S402 in TONE, monitor IC404-16 for tone pairs when any of the keys on keypad S405 are pressed. If there are no pulses at IC404-16, or an incorrect tone pair, suspect IC404 or keypad S405 or the wiring between the two components.

Next, check the IC404 external components, such as X401, C413, and C414. Make certain that IC404-2 is high. If low, suspect D408, Q407, Q401, R431, R411, and S402.

If there are tone pairs at IC404-16, check that the tone pairs are applied to the modulator and audio circuits through C411 and R333.

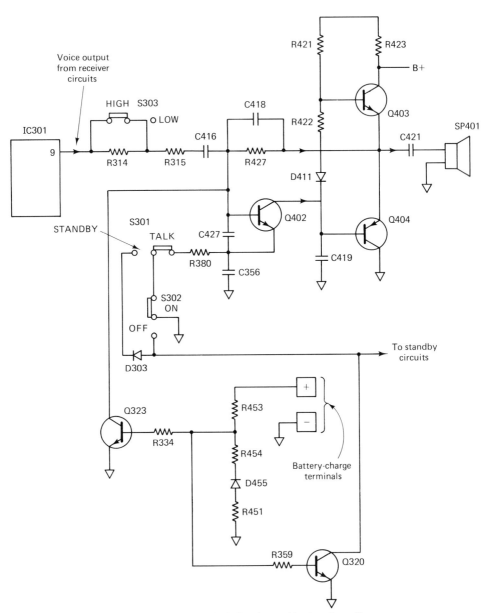

FIGURE 5-47. Tone-dial-circuit troubleshooting diagram

5-24. TELEPHONE RINGS BUT NO SOUND ON PORTABLE-UNIT SPEAKER

If the telephone rings, but a voice cannot be heard on the portable-unit earpiece speaker (with an assumed-good base unit; Sec. 5-11) the problem is likely to be in the portable-unit audio amplifier circuits.

Figure 5-48 shows the circuits involved. If the telephone rings, it is reasonable to assume that the portable-unit receiver circuits are operating properly and

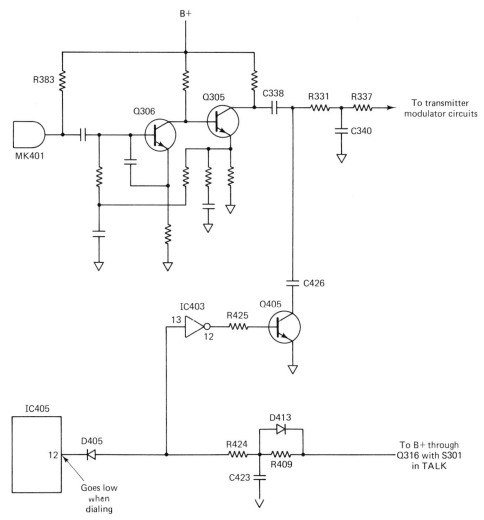

FIGURE 5-48. Portable-unit audio-circuit troubleshooting diagram

that there is an output available at IC301-9. Trace this voice output through the receiver audio circuits to speaker SP401 using an oscilloscope or other audio signal tracer.

As an example, if audio is available at IC301-9 but not at the base of Q402, suspect R314, R315, and C416. Also note the setting of HIGH/LOW switch S303.

Before you trace audio from point to point through the circuits, check that the collector of Q323 is not at ground. Q323 is an audio-mute transistor used to cut off the receiver audio circuits during *automatic standby*.

As discussed in Sec. 5-27, Q323 is turned on by the voltage across R453 and R454 when the portable-unit battery is charging. (Q320 is also turned on to produce the same conditions as when the STANDBY/TALK switch S301 is set to STANDBY.) Q323 is turned off when the portable-unit battery is not charging.

With Q323 on, the receiver audio is shorted to ground. The collector of Q323 is at (or near) zero volts with Q323 on. When Q323 is turned off (battery not charging), the collector rises to about 1 V.

Also note that the emitter of audio amplifier Q402 is returned to ground through S301 (in TALK), S302 (at ON), and R380. Typically, the collector of Q402 is at (or near) zero volts when S301 is in TALK and rises to about 2.5 V when S301 is moved to STANDBY.

5-25. TELEPHONE RINGS BUT VOICE FROM PORTABLE UNIT CANNOT BE HEARD

The procedures described thus far establish that the portable unit is not dead, that the telephone line can be seized, and that the telephone rings when called or a test ring is applied. These procedures prove that the transmitter and receiver of both the portable unit and base unit are probably good. So if voice cannot be heard on another telephone (with an assumed-good base unit; Sec. 5-10), the problem is likely in the portable-unit voice modulation circuits.

Figure 5-49 shows the circuits involved.

With the telephone turned on (TALK), speak into the portable-unit microphone MK401 (or inject an audio signal at the microphone input) and trace audio through to the junction of C338 and R331. Keep the following in mind when tracing audio through the circuits of Fig. 5-49.

First, make sure that microphone MK401 is receiving B+ operating voltage through R383. Current must flow through MK401 to produce an audio output.

Next, make sure that Q405 is not at ground (turned on). Q405 is an audio-mute transistor used to cut off the transmitter audio circuits during standby and when dial pulses are present.

Q405 is turned on when Q316 is turned off during standby. With Q316 off, B+ is removed and pin 13 of IC403 goes low. This low is inverted to a high at IC403-12 and causes Q405 to turn on. With Q405 on, the microphone audio is

Telephone Rings But Voice From Portable Unit Cannot Be Heard 177

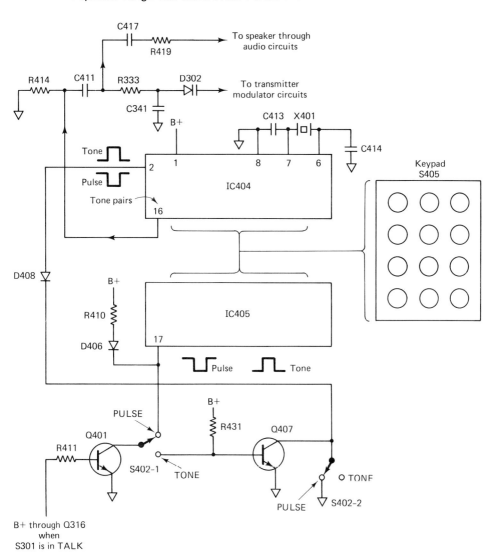

FIGURE 5-49. Portable-unit microphone-circuit troubleshooting diagram

shorted to ground and does not reach the transmitter circuits. Q405 is also turned on by dial-pulse generator IC405 when dialing (with a low at IC405-12).

When there is no dialing and S301 is set to TALK, Q316 turns on and connects B+ (high) to IC403-13. This high is inverted to a low at IC403-12 and turns Q405 off. With Q405 off, the microphone audio is passed to the modulator circuits.

5-26. FLASH (CALL-WAITING) NOT OPERATIVE

If the telephone is operating normally in all other respects but the flash or call-waiting function is inoperative, the problem is most likely to be in the circuits of Fig. 5-26.

As discussed in Sec. 5-4.6, the flash circuits interrupt the 5.3-kHz pilot tone to the base unit momentarily (less than 1 second). This opens and closes the base-unit telephone line relay (rapidly) to signal the telephone exchange that you are ready to receive a second call (and put the first call on hold).

If the timing is incorrect for flash operation (relay opening and closing is too short to signal the telephone exchange, or if the circuits break contact with the telephone line so that the first call is lost), check the adjustment of VR301 as discussed in Sec. 5-5.5.

If the flash circuits are totally inoperative, check the wiring from the keypad FLASH (#) contact to IC303-3. Then check pins 1 through 6 of IC303. Pin 2 of IC303 should go low for approximately 400 to 800 ms when FLASH (#) is pressed. Also check diode D308.

5-27. AUTOMATIC STANDBY FUNCTION NOT OPERATIVE

Figure 5-50 shows the circuits involved. The portable unit is placed in the standby condition automatically when the battery is being charged. When the battery voltage drops below a certain level, charging current from the base-unit power supply (or external source at J406) flows through D451.

When the portable unit is on the base-unit cradle and the charging terminals are making contact, current flows through R453, R454, and D455. The voltage at the junction of R453 and R454 turns Q320 on. This connects the junction of R363 and R374 to ground (through Q320) and has the same effect as setting STANDBY/TALK switch S301 to STANDBY (as discussed in Sec. 5-22).

When the portable unit is removed from the cradle and the charging terminals are disconnected, there is no voltage at R453/R454 and Q320 turns off, restoring control of the standby/talk circuits to S301.

If the automatic standby circuits appear to be inoperative, check the collector of Q320 with S301 in both TALK and STANDBY, with the portable unit on the cradle (charging terminals connected). There should be substantially no change of Q320 collector voltage in either position of S301. The collector of Q320 should remain at about 0.5 V.

Now check the collector of Q320 with S301 in TALK and the portable unit lifted from the cradle (charging terminals disconnected). The collector should rise to about 3 V (with S301 in TALK) and drop to 0.5 V (with S301 in STANDBY).

If these conditions are not correct, check Q320, R453, R454, and D455. Of course, make certain that there is a charging voltage across the charging terminals.

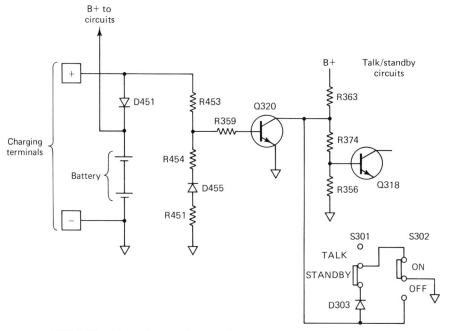

FIGURE 5-50. Automatic-standby-circuit troubleshooting diagram

Also note that the automatic standby function does not operate when the battery is charged from an external source through J406.

Similarly, the automatic standby voltage applied to Q320 is also applied to the audio-mute transistor Q323 as described in Sec. 5-24 and shown in Fig. 5-48.

5-28. NO KEYSTROKE PULSE GENERATED WHEN DIAL KEYS ARE PRESSED

Figure 5-51 shows the circuits involved. When the keypad buttons are pressed, the 2.5-kHz paging oscillator tone is heard in the earpiece speaker SP401. This is the same 2.5-kHz signal that is reproduced on the call speaker SP402 (shown in Fig. 5-21) during ring/call operation.

The anode of D410 is connected to B+ through R412. When call signals are present, Q314 shorts the anode of D410 to ground, reverse biasing D410. This turns on the 2.5-kHz oscillator IC403, which is heard on the call and earpiece speakers.

Diode D410 is also reverse-biased (anode shorted to ground) when the keypad buttons are pressed. This turns on the IC403 oscillator, which is heard on the earpiece speaker SP401 through the audio amplifier circuits.

The 2.5-kHz signal is not heard on the ring/call speaker SP402 (when the keypad buttons are pressed) because D305 is reverse biased when there is no valid

180   Cordless Telephone Troubleshooting

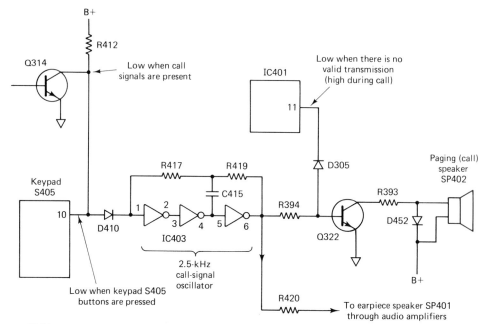

FIGURE 5-51. Keystroke-tone-circuit troubleshooting diagram

transmission signal and IC401-11 goes low. IC401-11 is high only during a ring/call condition (when the base-unit CALL button is pressed or when there is a ring signal on the telephone line input to the base unit).

If there are no keystroke signals on the earpiece speaker SP401 when the key-pad buttons are pressed, first check that the ring/call signal is heard on the ring/call speaker SP402 during a call. If not, refer to Sec. 5-21. If so, the 2.5-kHz oscillator IC403 can be presumed good.

Next, check that voice is heard on the earpiece speaker SP401. If not, refer to Sec. 5-24. If so, the audio amplifier circuits can be presumed good.

If both the audio amplifier circuits and the 2.5-kHz oscillator are good, the problem is most likely to be in the keypad S405 itself or possibly in the wiring between the keypad and D410.

MODEM AND TELEPHONE INTERFACE TROUBLESHOOTING

This chapter describes a series of troubleshooting and service notes for some typical modems (*mo*dulator-*dem*odulator) and telephone interface units (also known as *acoustic couplers*). As in the case of cordless telephones described in Chapter 5, it is not practical to provide specific troubleshooting procedures for every modem and interface. Instead, we describe typical examples.

It is assumed that you are already familiar with digital electronics as used in computers and the basics of data communications (how computers communicate over telephone lines). If not, your attention is invited to the author's best-selling *Handbook of Data Communications* (Englewood Cliffs, N.J. Prentice-Hall, Inc., 1984).

Before we get into modem and interface circuits let us review modem basics.

6-1. BASIC MODEM SYSTEM

Figure 6-1 shows a typical modem system application together with the standard system frequency assignments. As shown, a modem fills the need in a data communications network to provide interface between a telephone network, which carries analog information, and a computer system that operates on digital information. Note that both the transmitting (modulation) and receiving (demodulation) functions are contained within each modem, permitting *duplex* operation (transmission and reception at both ends of the telephone lines, as in the case of conventional telephone operation).

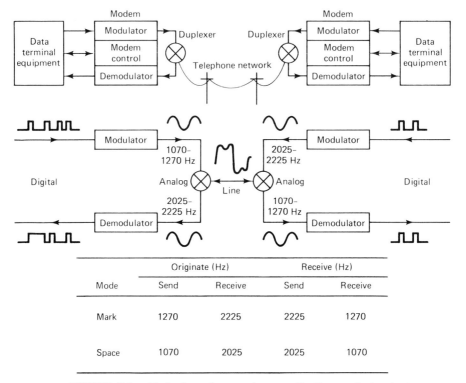

FIGURE 6-1. Typical modem system application and standard system frequency assignments

Basically, the modem converts logical 1 and 0 levels into audio-frequency tones and back again to 1s and 0s. These tones have the specific frequencies listed in Fig. 6-1. Two pairs of tones are listed for each modem, one set for transmitting and one set for receiving, so that simultaneous two-way (full-duplex) operation is possible over a single transmission channel.

The terminal modem that places a call is referred to as the *originate modem* (transmitting tones of 1070 and 1270 Hz), whereas the terminal receiving this call is the *answer modem* (transmitting tones of 2025 and 2225 Hz).

6-1.1 RS232C and CCITT V.24

The frequencies shown in Fig. 6-1 are part of EIA (Electronics Industries Association) RS232C, which is the most widespread and universal standard for communications between computers and computer-related equipment in the United States. RS232C is generally compatible with CCITT (International Consultative Committee for Telegraphy and Telephony) V.24, which is a commonly used international standard.

Figure 6-2 shows the pin or line format for RS232C. A total of 25 lines are

Basic Modem System 183

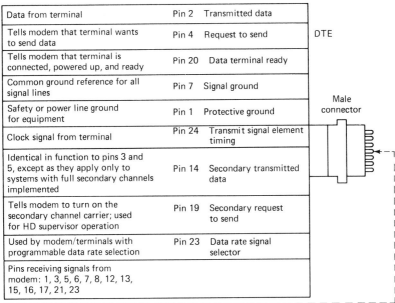

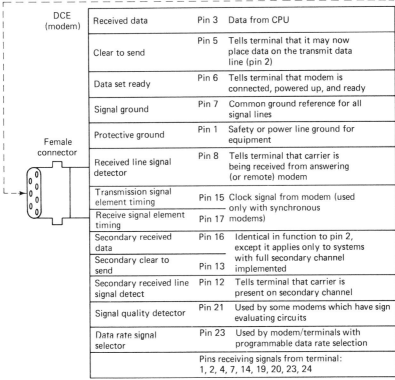

FIGURE 6-2. Basic pin or line format for RS232C

184 Modem and Telephone Interface Troubleshooting

available between the data terminal or computer and the modem. Note that all 25 lines are not necessarily used in all data communications networks. However, when a line is used, the format or function is the same for all equipment covered by RS232C. For example, if pin 8 of a data terminal is connected to pin 8 of a modem (as it is in most cases), the signal on that line tells the terminal that a carrier signal is being received from the answering (or remote) modem over the telephone lines by the calling modem.

Figure 6-3 lists the basic characteristics for RS232C as well as RS422 and

Parameter	RS232	RS422	RS423
Line length (recommended max may be exceeded with proper design)	50 ft	1200 m (4000 ft)	1200 m (4000 ft)
Input impedance, Z	3-7 kΩ 2500 pF	$>$ 4 kΩ	$>$ 4 kΩ
Max frequency (baud)	20 kbaud	10 Mbaud	100 kbaud
Transition time* (time in undefined area between 1 and 0) tr = 10–90%	4% of t or 1 ms	tr $\leqslant$ 0.1 t t $\geqslant$ 200 ns tr $\leqslant$ 20 ns t $<$ 200 ns	tr $\leqslant$ 0.3 t t $<$ 1 ms tr $\leqslant$ 300 μs t $>$ 1 ms
dV/dt (waveshaping)	30 V/μs	Dependent on cable length	Dependent on cable length
Mark (data 1) Space (data 0)	$-$3 V +3 V	A $<$ B A $>$ B	A = negative B = positive
Common-mode voltage (for balanced receiver)	—	$-$7 V $<$ V_{CM} $<$ +7 V	—
Output impedance, Z	—	$<$ 100 balanced	$<$ 50 Ω
Open-circuit output voltage, V_0	3 V $<$ $\mid V_0 \mid$ $<$ 25 V	$\mid V_0 \mid$ $\leqslant$ 6 V[†]	4 V $\leqslant$ $\mid V_0 \mid$ $\leqslant$ 6 V
V_t = loaded V_0	5 $<$ $\mid V_0 \mid$ $<$ 15 V 3 to 7-kΩ load	2 V or 0.5 V_0 $<$ $\mid V_t \mid$ [††] 100-Ω balanced load	$\mid V_t \mid$ $\geqslant$ 0.9 $\mid V_0 \mid$ 450-Ω load
Short-circuit current	500 mA	150 mA	150 mA
Power-off leakage (V_0 applied to unpowered device)	$>$ 300 μA 2 V $<$ $\mid V_0 \mid$ $<$ 25 V V_0 applied	$<$ 100 μA 0 V $<$ V_0 $<$ 6 V V_0 applied	$<$ 100 μA $\mid V_0 \mid$ $<$ 6 V V_0 applied
Min receiver input for proper V_0	±3 V	200-mV differential	200-mV differential

*t is bit period.
[†] Across output, or output to ground.
[††] Whichever is greater.

FIGURE 6-3. Basic characteristics of RS232C, RS422, and RS423

RS423. Note that RS422 and RS423 are newer standards used when communications lines are very long or where data bits are transmitted at very high speeds. In this book we concentrate on RS232C since the standard is used by most present-day communications systems in the United States.

6-1.2 DTE and DCE

RS232C defines all equipment as either *data terminal equipment* (DTE) or *data communications equipment* (DCE). The video terminal (keyboard and cathode-ray tube display) is one of the most popular DTE devices. The modem and acoustic coupler (telephone interface) are the most popular DCE devices. Note that some video terminals include a modem. Such instruments are therefore both DCE and DTE.

6-1.3 Typical Serial Data Transmission System

The following paragraphs provide a brief description of a typical serial data transmission system, which is used with the Motorola 6800-series microprocessor systems. Two devices are involved, the MC6850 ACIA (asynchronous communications interface adapter) and the MC6860 digital modem.

ACIA. Figure 6-4 shows the interfacing relationship to the system, and basic block diagram, of the ACIA. As shown, the ACIA provides the data formatting and control to interface asynchronous serial data between the computer and modem. The parallel data of the M6800 bus system is serially transmitted and received (full duplex) by the ACIA, with proper formatting and error checking. The term "full duplex" applied here means that the serial data bits can be transmitted and received simultaneously on the same telephone line.

The functional configuration of the ACIA is programmed via the data bus (in the computer) during system initialization. A programmable control register provides variable word lengths, clock division ratios, transmit control, receive control, and interrupt control. Three input/output (I/O) lines are provided to control external peripherals or modems. A status register is available to the computer and reflects the current status of the transmitter and receiver.

Digital modem. Figure 6-5 shows the relationship to the system, the block diagram, and a typical system interface for the digital modem. The modem is essentially a subsystem designed to be integrated into a wide range of equipment using serial data communications. The modem provides the necessary modulation, demodulation, and supervisory control functions to implement a serial data communications link, over a telephone line (or other voice-grate channel), using FSK (frequency-shift keying) at bit rates up to 600 bits per second (bps). The modem can be implemented into a wide range of data storage devices, remote data communciations terminals, and I/O devices.

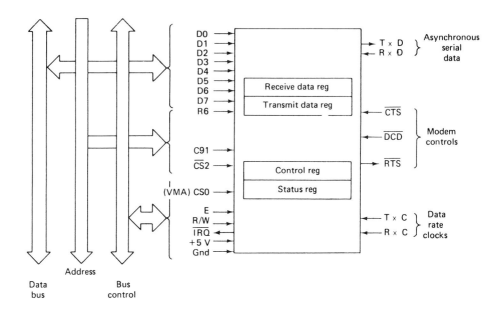
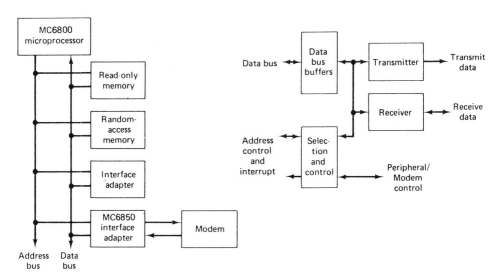

FIGURE 6-4. MC6850 ACIA, interfacing, relationship to system, and basic block diagram

Basic Modem System

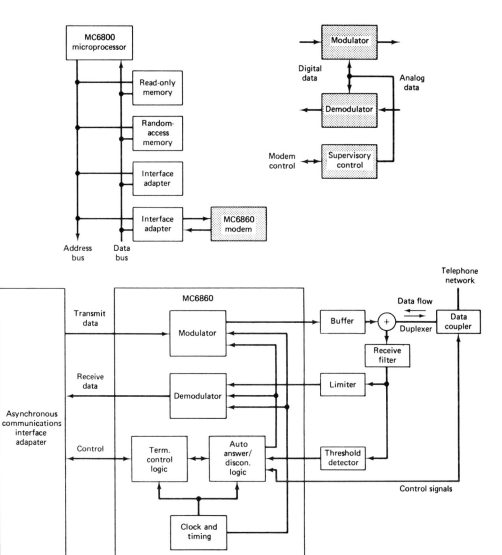

FIGURE 6-5. MC6860 digital modem, relationship to system, basic block diagram, and typical system configuration

6-1.4 Teletype Interface

Existing teletype (TTY) lines can be used for serial data transmission, although telephone lines are far more popular in present-day digital communications. When TTY is used, the interface is called a *20-mA current loop* or simply a *current loop*.

6-1.5 Marks and Spaces

The two logical states for serial data transmission systems are called a *mark* and a *space*. As shown in Fig. 6-3, for RS232C, a mark (or logic 1) is a minus voltage, whereas a space (logic 0) is a positive voltage. For TTY, a mark is 20 mA flowing in the current loop between the two devices, whereas a space is zero current in the 20-mA current loop.

While on the subject of marks and spaces, in conventional TTL (transistor–transistor logic) equipment such as that found in computers and terminals, a mark is a logic 1 (which can be positive or negative), whereas a space is a logic 0.

Figure 6-6 shows a typical serial data transmission format (the character "M" in ASCII, the American Standard Code for Information Interchange).

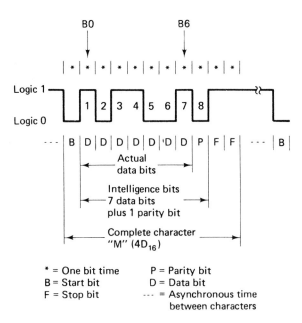

FIGURE 6-6. Serial data transmission format (showing character "M" in ASCII)

6-2. MODEM I TROUBLESHOOTING

This section describes troubleshooting for a typical telephone modem. The Radio Shack Modem I is selected as an example. We start with a description of the modem circuits. It is essential that you understand operation of any modem (or other electronic device) before you attempt troubleshooting!

6-2.1 Circuit Descriptions

The Modem I is a device that allows two computer terminals to communicate via standard telephone lines. Figure 6-7 is a block diagram of the complete modem circuit.

Power supply. Power for the modem is derived from a low-voltage a-c adapter with an output of 15 V, as shown in Fig. 6-8.

The a-c voltage is half-wave rectified by CR2/CR3 and filtered by C9/C14 to provide the unregulated d-c voltage. The +12-V supply is provided by a three-terminal regulator U1. The −12-V supply is provided by a zener regulator consisting of R3, CR1, and CR4. The +5-V supply is provided by a three-terminal regulator U9 which receives input from the +12-V regulated supply.

Telephone line interface. The telephone line is matched to the modem by the circuit shown in Fig. 6-9. The matching circuit consists of a 600:600-Ω hybrid transformer T1 with a tapped secondary. A balancing network R37 and

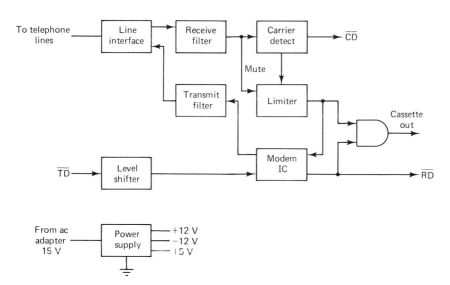

FIGURE 6-7. Radio Shack Modem I block diagram

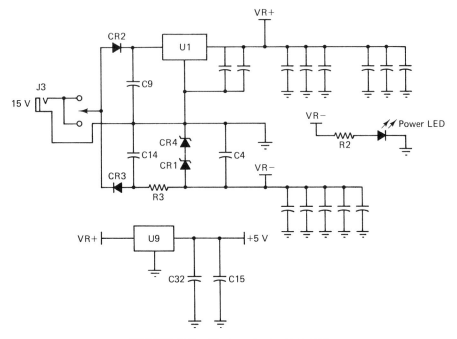

FIGURE 6-8. Modem I power supply

C16 is used to isolate the transmitted tones (at T1, pin 8) from the received tones (T1, pin 5). Termination is provided for the T1 secondary by R26 and R28. To protect the modem from voltage spikes on the telephone line, a 22-V MOV (metal-oxide varistor) ZN1 is located across the T1 primary.

A word of caution! If you must probe the ring detector circuits of a modem during troubleshooting, be sure that the test equipment is plugged into an isolation transformer. The voltages involved can damage either the modem or the telephone lines (or both) if the test equipment is not isolated from ground.

Receive filter. The receive filter U6a, U6b, and U2b shown in Fig. 6-9 removes signals of all undesired frequencies from the incoming signal. The filter is a six-pole Chebyshev with 1-dB ripple in the passband. The filter characteristics are switchable for answer and originate modes (as controlled by MODE switch S1).

In the originate mode, the center frequency is approximately 2100 Hz; in the answer mode, the center frequency is approximately 1170 Hz. The bandwidth of the filter in either case is approximately 300 Hz. An amplifier U2a precedes the filter. U2a has some filtering to remove the 1-MHz clock noise generated by the crystal. Filter output is fed to the carrier detection and limiting circuits.

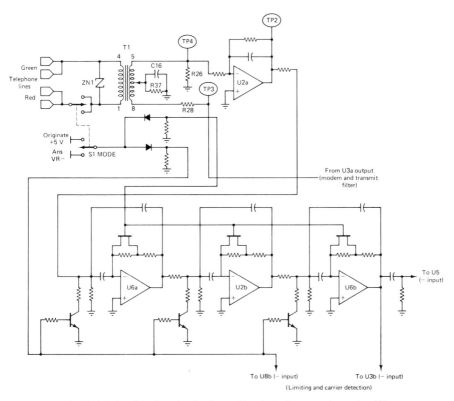

FIGURE 6-9. Modem I telephone line interface and receive filter

Limiting and carrier detection. The modem IC U4 requires a 50 ± 4% square-wave signal as receive data input. Squaring of the incoming signal is done by the high-speed comparator U5 shown in Fig. 6-10.

The sine-wave signals from the receive filter are a-c coupled by C23 to the inverting input of the comparator to remove any d-c offset. The comparator toggles on successive zero crossings of the input signal.

The limiter has an open-collector output which is wire-ANDed with the output of the carrier-detection circuit. This allows input to the modem IC only while a valid carrier is being received.

Carrier detection is done via an active filter U3 and comparator U8a/U8b. One section of comparator U8b is connected to the MODE switch S1 to compensate for gain mismatch between the answer and originate filters.

The tone from the filter is half-wave rectified by CR7/CR8 and filtered by C30. The tone is then compared to the threshold. When the level at the minus (−) input of U8a exceeds the level at the plus (+) input, the output is pulled low. This turns off transistor Q8 and allows the limiter to pass tones to the modem IC.

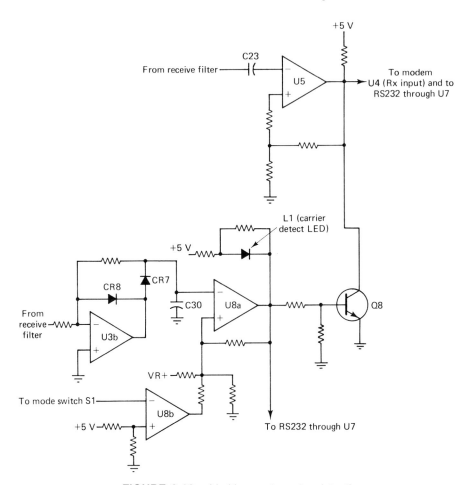

FIGURE 6-10. Limiting and carrier detection

Modem and transmit filter circuits. The modulation and demodulation of the FSK data are done internally by the modem IC U4 shown in Fig. 6-11. U4 is a CMOS IC, clocked by the 1-mHz crystal Y1. The modem IC generates a particular frequency, depending on the MODE switch S1 setting and the input data. The modem IC also detects the square-wave signal from the limiter and generates logic 1s and 0s for transmission to the computer terminal.

The transmit filter U3a removes the higher harmonics from the signals generated by the modem IC. This prevents "singing" for feedback between the transmit and receive filters. The telephone hybrid transformer T1 also acts as a filter for the transmitted tones.

The transmit power or level is set by potentiometer R1 on the input of the transmit filter. Note that R1 is the only adjustment control for the entire modem circuit.

Data I/O drivers and receivers. Communications to and from the computer terminal is usually in RS232C format (Figs. 6-2 and 6-3). The modem includes drivers and receivers necessary to implement the RS232C functions as shown in Fig. 6-12.

The drivers U7 accept TTL inputs (from 0 to 5 V) and produce RS232C-compatible outputs. Capacitor C28 on the data output line is used to limit the rise and fall times of the digital data. The receiver U8c accepts RS232C-compatible data and produces a TTL output.

On those computers with cassette storage, the modem receiver output can be applied to the cassette input in the form of tone bursts. This is controlled by RS232/cassette switch S2.

With S2 in RS232, the receiver output from U4 is applied directly to computer terminal port BB and to the cassette port. With S2 in Cassette, the receiver output from U4 is applied to the cassette port through U8d and U7. This changes the receive threshold from 3 V to 0.6 V. Also, with S2 in Cassette, the input to U8c is shorted to ground through CR10, preventing transmit information from passing to U4 through U8c.

Cassette data bits to the computer terminal are generated by comparator U8d and driver U7. The comparator forms an inverter for the normal RS232C

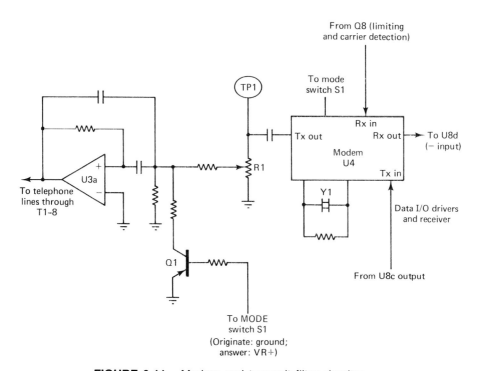

FIGURE 6-11. Modem and transmit filter circuitry

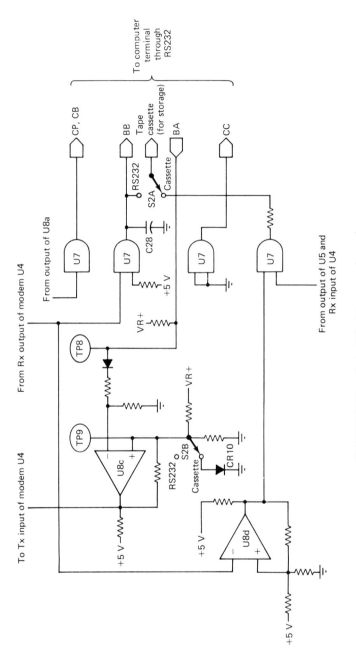

FIGURE 6-12. Data I/O drivers and receivers

data bits. The resultant inverted bits are ANDed in the computer with the normal modem output to produce tone bursts.

The resistor at the output of the driver limits the voltage swing to the cassette port by forming a divider with the internal resistor termination of the computer.

6-2.2 Adjustments

The only adjustment required for the modem involves setting the level of the transmit output from modem U4 to the telephone line. Figure 6-13 shows the adjustment diagram.

With a 600-Ω resistor connected across the hybrid transformer T1, monitor the ring-to-tip output. Adjust R1 for an output of about 0.2 V(rms) (-10 dB), with a steady digital input at J1-2, as shown in Fig. 6-13. As discussed in Sec. 6-2.3, this adjustment should produce an output of about 1.8 V peak to peak at pin 7 of U3a.

6-2.3 Troubleshooting

Because a modem performs its function automatically, troubleshooting on the basis of trouble symptoms is often impractical. (Many modems have no operating controls or indicators.) Instead, modem manufacturers often recommend that the

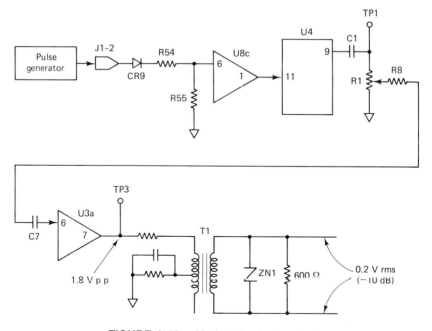

FIGURE 6-13. Modem I adjustment diagram

waveforms produced by individual circuits be checked. That is the approach we use here.

However, there are certain general troubleshooting procedures that can help isolate trouble to a section of the modem and eliminate unnecessary checks. For example, if the modem does not answer the phone but can originate calls, it is useless to check for filter problems, but it is quite logical to check the ring detection circuits or the modem microprocessor.

Figure 6-14 summarizes these check procedures for any modem. The following paragraphs then go on to describe the checks of individual circuits, based on waveform analysis.

Transmitter output. Figure 6-15 shows the circuits and waveforms involved for the transmitter output section.

The waveform shown is typical of a modem microprocessor output. The waveform is a digitally produced sine wave that can have a frequency of 1070 Hz or 1270 Hz (when the modem is in the originate mode) and a frequency of 2025 Hz or 2225 Hz (when in the answer mode).

If there is no output at U4-9, make certain that S2 is in RS232 (not in Cassette) and that there are digital pulses at J1-2. If there are no pulses at J1-2, the problem is in the computer or terminal, not in the modem.

If there are pulses at J1-2, check for pulses at U4-11 and U8-6. If there are pulses at U8-6 but not at U4-11, suspect U8, R52, R53, R61, and R62. If pulses are absent at U8-6, suspect CR9, R54, R43, and R55.

If there are pulses at U4-11 but not at U4-9, suspect U4, Y1, R5, and the wiring to U4.

Transmit filter. Figure 6-16 shows the circuits and waveforms involved for the transmit filter section. The waveform shows the result of the transmit filter on the digitally produced sine wave. As discussed in Sec. 6-2.2, the transmit level is adjusted by R1 to -10 dB, measured at the telephone line with a 600-Ω resistor across the tip and ring of connectors to the T1 transformer primary (J4-3 to J4-4). This adjustment produces an output from filter U3a of about 1.8 peak to peak.

Trouble Symptom	Check Procedures
Modem dead	Power supply, processor
Modem does not answer	Ring detector circuits, processor, modem IC carrier-detect
No tone dial	Tone-dial IC, processor, hybrid transformer
No pulse dial	Pulse-dial relay, processor
Modem does not program	Modem controls, RS232 drivers, processor
Erratic performance	Filter response, limiter, carrier-detect, modem IC

FIGURE 6-14. Summary of modem troubleshooting check procedures

If the waveform at U3-7 is not normal (as shown in Fig. 6-16), try adjusting R1 first. If the problem cannot be corrected by adjustment, check for a waveform at U3-6. (This waveform should be approximately the same as that shown for the output of U4-9 in Fig. 6-15, except of lower amplitude.)

If the waveform is absent or abnormal at U3-6, suspect C1, C7, C8, R27, R1, R8, R7, R9, R6, or Q1. Note that Q1 is turned on through Slc in the answer mode. This shorts the transmit signals from U4-9 to ground through Q1. In the originate mode, Q1 is off since the base of Q1 is connected to ground through R6 and Slc.

If the waveform is normal at U3-7 but not at the output to the telephone line (T1 primary), suspect T1, R28, and S1. It is also possible that C16, ZN1, or R37 can be at fault.

Receive filter input. Figure 6-17 shows the circuits and waveforms involved for the receive filter input section.

The waveform from transformer T1 contains many harmonics which are mixed as shown in Fig. 6-17. (The waveform of Fig. 6-17 is typical of incoming FSK data.) The signal is first amplified by U2a and then goes on for further amplification and filtering.

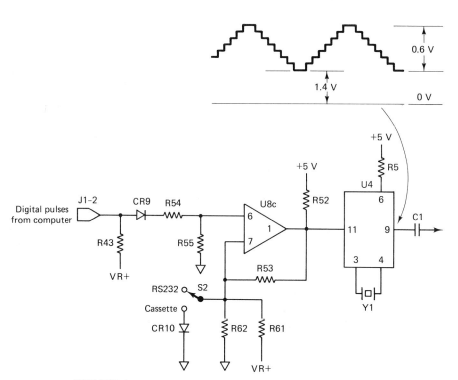

FIGURE 6-15. Transmitter-output troubleshooting diagram

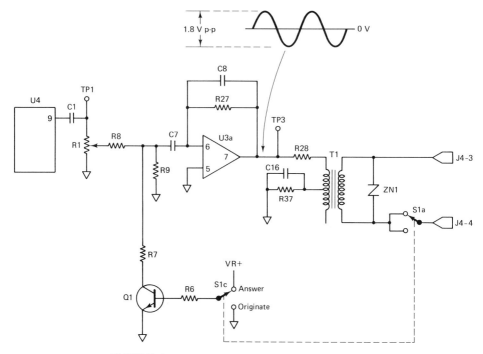

FIGURE 6-16. Transmit filter troubleshooting diagram

Note that the bandpass characteristics of the receive filter input circuits are changed when S1 is moved between originate and answer. So if the problem is one of good operation in answer but not in originate (or vice versa), make sure that S1 is set properly.

Trace the incoming signals from the telephone line through to the output at U6-7, using the waveforms of Fig. 6-17 as a guide. This should show up any problems in the receive filter input circuits.

Receive filter output. Figure 6-18 shows the circuits and waveforms involved for the receive filter output section. The last two stages of the receive filter remove the harmonics from the waveform as shown in Fig. 6-18. (Compare the waveforms of Figs. 6-17 and 6-18.) The last two stages also increase the amplitude of the receive signal.

Note that the bandpass characteristics of the receive filter output circuits are also changed when S1 is moved between originate and answer.

Trace the incoming signals from the receive filter input circuits through to the output at U6-1, using the waveforms of Fig. 6-18 as a guide. This should show up any problems in the receive filter output circuits.

Limiting and carrier detection. Figure 6-19 shows the circuits and waveforms involved for the limiting and carrier detection section. The waveform at U5-7 shows the limiting action produced by U5. The signal at U5-7 is applied

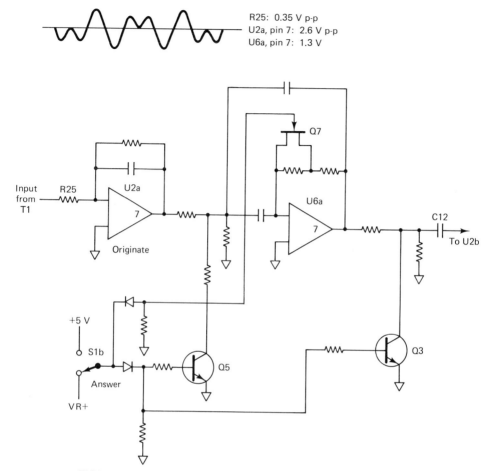

FIGURE 6-17. Receive-filter-input troubleshooting diagram

to U4 as the receive input at U4-1. This input is cut off when Q8 is turned on by a signal from U8-2. When U8-2 is high, Q8 is turned on; the output at U5-7 is shorted to ground and does not reach U4-1. When U8-2 is low, Q8 is turned off, and the signals from U5-7 are passed to U4-1. Also, when U8-2 is low, CARRIER DETECT indicator L1 (and LED) turns on to indicate that there is a valid carrier being received.

The input to the carrier-detect circuit is at U3-2. (This same input is also applied to U5 through C23 and comes from the receive filter output at U6-1.) The waveform at U3-1 shows the action of the half wave rectifier CR7/CR8. Note what when CR7 conducts, the input signal is amplified and squared off at the top. When CR8 conducts, gain resistor R16 is bypassed and produces almost a 0.5-V swing at the output.

The rectified output from CR7/CR8 is filtered by C30, producing a voltage at U8-4. The output at U8-2 goes low when the voltage exceeds the threshold set by comparator U8b and resistors R50, R59, and R60. This turns on CARRIER DETECT indicator L1, turns Q8 off, and permits passage of signals received on the telephone line to reach the computer.

If the problem is one of good transmission but no reception, try injecting a signal at the primary of T1 (Fig. 6-17) and check that the CARRIER DETECT indicator L1 turns on. Use a test signal of 1250 Hz, with an amplitude sufficient to produce a signal of 0.35 V peak to peak at R25 (Fig. 6-17), for injection at the primary of T1. Also check that there is a signal at the output to the computer (U7-11 or J1-3).

If L1 turns on but there is no signal at U7-11 or J1-3, suspect U5, U4, or U7. The fact that L1 turns on indicates that there is a valid carrier at U6-1 and that the carrier detection circuits are operating normally.

If L1 does not turn on, make sure there is a proper signal at U6-1. (As shown in Fig. 6-18, the signal should be a sine wave with an amplitude of about 6.5 V.) If not, suspect the receive filter input and/or output circuits.

If there is a proper signal at U6-1 but L1 does not turn on, suspect L, U3, U8, CR7, CR8, and the related circuit components.

If L1 is turned on, check the collector of Q8. The collector should be at some voltage above ground. If not, suspect R44, R45, R63, or Q8.

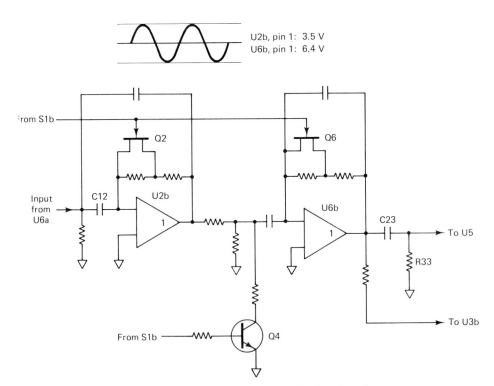

FIGURE 6-18. Receive-filter-output troubleshooting diagram

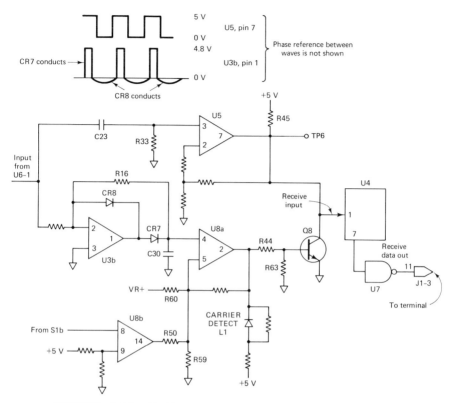

FIGURE 6-19. Limiting and carrier-detection-circuit troubleshooting diagram

6-3. MODEM II TROUBLESHOOTING

This section describes troubleshooting for another telephone modem. The Radio Shack Modem II is selected as the example. Again, we start with a description of the modem circuits.

6-3.1 Circuit Descriptions

The Modem II is a direct-connected FSK modem which can automatically dial and automatically answer. The Modem II incorporates laser-trimmed filter networks and a custom microprocessor to handle modem operation. The microprocessor can be programmed by sending ASCII data (bit strings) to the modem via the RS232C lines.

Figure 6-20 is a block diagram of the complete modem circuit. The modem connects to the telephone line using standard modular plugs and jacks. The telephone line is monitored by the ring detect circuitry to determine if a call is

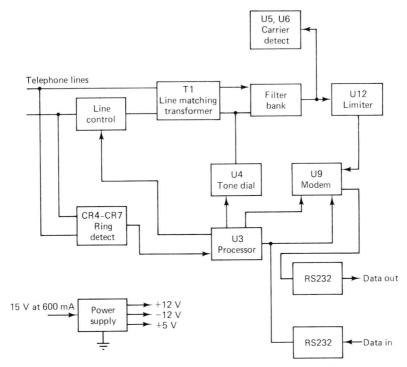

FIGURE 6-20. Radio Shack Modem II block diagram

coming in. A hybrid transformer T1 is used to match the impedance of the telephone line.

The FSK tones are first filtered and then limited to produce a 50% duty cycle square wave at the limiter output. The square-wave signals pass to the CMOS modem IC U9, which demodulates the data and sends the data to the computer over RS232C lines. Data bits from the computer are examined by the processor U3 and then sent to the modem IC U9, where the bits are converted into sine-wave form. The sine-wave signals are filtered and sent to the telephone lines.

Power supply. Power for the modem is derived from a wall-mounted transformer which produces 15 V ac at 600 mA. This low-voltage ac is full-wave rectified by dioide bridge U13, as shown in Fig. 6-21.

The positive half-cycle is filtered by C30 and regulated to 12 V by VR3. A second voltage regulator VR2 with series dropping resistor R49 produce a +5-V regulated voltage. The negative half-cycle is routed to a voltage doubler, consisting of diodes CR13/CR14 and capacitors C29/C37. This output is regulated to −12 V by VR1. Power for multiplexers U11 and U17 is provided by the +12-V and −12-V supply through 7.5-V zener diodes CR10 and CR11, with series-limiting resistors R43/R52.

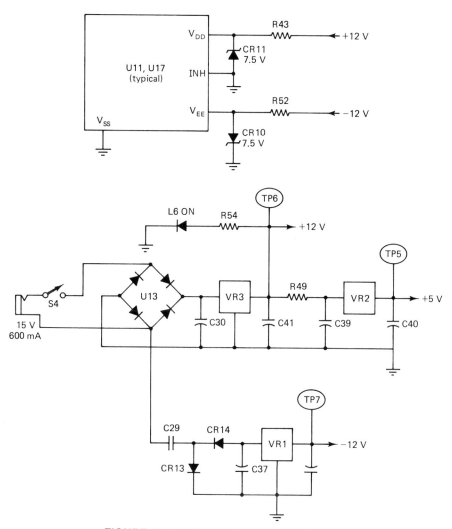

FIGURE 6-21. Modem II power supply

Ring detector circuits. In the auto-answer mode, the modem monitors the telephone line for a ring voltage. During a ring, a sine-wave voltage of approximately 80 V is present on the telephone line. Also, the telephone lines are approximately 48 V with respect to the modem. For this reason, an opto-isolator U8 is used to isolate the modem from the telephone line, as shown in Fig. 6-22.

The incoming ring voltage is applied to the full-wave bridge rectifier CR4–CR7 through impedance-matching circuit R31/C53, which approximates a telephone "bell" for the telephone line. The rectified ring voltage is slightly filtered by C12 and applied to a voltage switch formed by Q2 and Q3. When the voltage

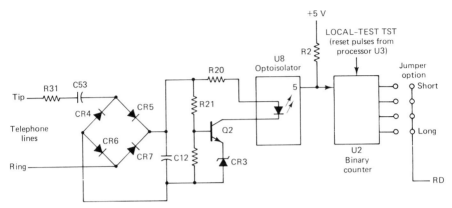

FIGURE 6-22. Modem ring-detect circuitry

on the base of Q2 reaches V3 (2.4 V from the zener diode and +0.6 V from the drop in Q2) the transistor conducts, pulling current through R20 and turning on the opto-isolator U8.

Opto-isolator U8 has an open-collector output, pulled up by R2. The voltage at pin 5 of U8 is normally +5V and pulses to zero during ring.

These pulses clock the CMOS binary counter U2. A jumper option is provided to tap the output to determine when the processor U3 answers (U3 answers when the selected U2 counter output is high). Counter U2 is reset on power-up by processor U3 pulsing the LOCAL-TST line, which is connected to the U2 reset line. The reset line is also used to keep the modem from answering incoming calls during a local test.

Line control and matching. The modem connects to the telephone lines through standard RJ11C connectors and uses two relays K1 and K2 to control telephone line connections, as shown in Fig. 6-23. Relay K1 is used to pulse-dial the line, while K2 is used to seize the line.

Pulse dialing is accomplished by opening the contacts of relay K1 and placing an impedance-matching network, consisting of R44/C22, across the telephone line. Relay K1 is controlled by processor U3. A high-voltage driver U10 is used to convert the +5-V logic levels from the processor to operate the 12-V relays. When directed by processor U3, by asserting SEIZE, relay K2 closes and seizes the telephone line.

The line interface uses a 600:600-Ω transformer T1 with a center-tapped secondary. The secondary is connected to an impedance-balancing network composed of R24, C16, R32, and R40. Hybrid transformer T1 uses special construction techniques to achieve balance over a wide range of loop currents. ZN1 is located across the T1 primary to protect the modem from voltage spikes.

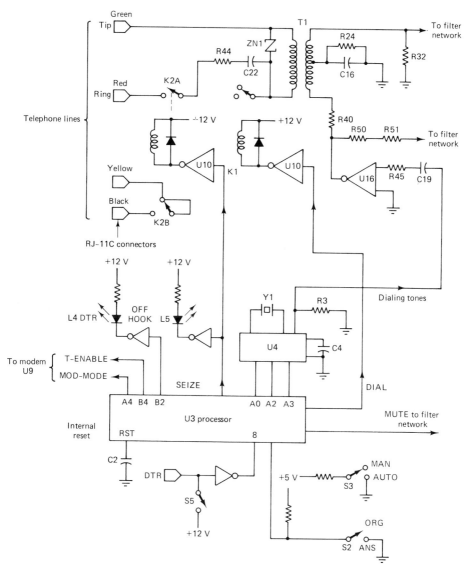

FIGURE 6-23. Modem II line control, matching, processor, and tone dialing

Filter network. The modem uses custom laser-trimmed resistor networks for the two filter sets. The low-band filter (center frequency 1170 Hz) uses two networks, while the high-band filter (center frequency 2125 Hz) uses three networks, as shown in Fig. 6-24.

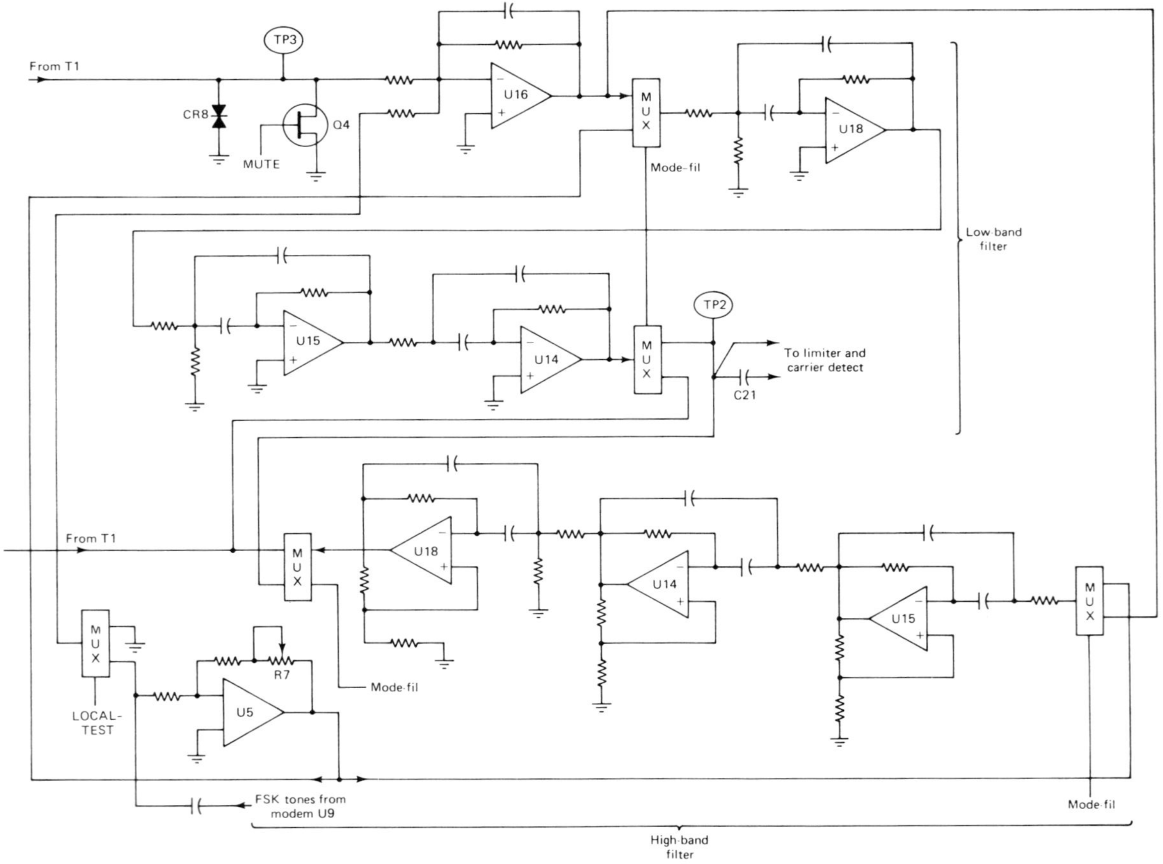

FIGURE 6-24. Modem II filter network

Both filters have 1-dB ripple Chebyshev bandpass responses with gains of approximately 16 dB. Most of the receive gain is in preamplifier U16. Transistor Q4 is used on the input to mute the incoming signal when the modem pulse-dials. Further limiting is provided by "click suppressor" CR8, which clips signals greater than 1.5 V peak to peak. Normal signals from another modem are around 100 mV at TP3.

The FSK tones from the modem IC U9 are attenuated by U5. Trimpot R7 is set for -10-dB output into a 600-Ω load. The tones are either sent to the transmitter filter or looped back through a CMOS multiplexer into the preamp U16. This is controlled by the status of the LOCAL-TEST circuit. The filtered tones are alternately mixed with the tone-dialing signals by the other half of U16 (Fig. 6-23). Of course, the two tones are never mixed with each other.

Limiter and carrier-detect circuit. The filtered FSK tones from the sending modem are pure sine waves at test point TP2, as shown in Fig. 6-25. The modem IC U9 requires a square wave for proper decoding. High-speed zero-

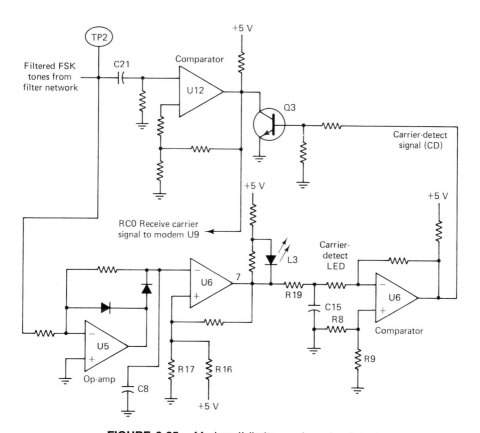

FIGURE 6-25. Modem II limiter and carrier-detect

crossing detector U12 is used to turn the sine waves into 5-V square waves. The output of U12 is open-collector, which is wire-ORed with transistor Q3. In turn Q3 receives a carrier-detect (CD) signal at the base. Unless there is a valid carrier, the output of comparator U12 is held low by Q3, preventing interference or undesired noise from being sent to the computer in the absence of a carrier on the telephone line.

The carrier-detect circuit consists of dual low-speed comparator U6 and half of dual op-amp U5. The op-amp is connected as a precision rectifier with a gain of 4. The sine-wave signal from the filter is half-wave rectified and then d-c averaged by C8. This d-c voltage is compared to a reference voltage set by the voltage divider, consisting of R16 and R17. When the voltage from the rectifier exceeds this reference, the output goes low, turning on carrier-detect LED L3.

Since there is a timing protocol for carrier-detect disconnect, a half-monostable circuit is used to time how long the carrier is present. This is done by charging C15 through R19 when pin 7 of U6 goes low. The time constant is about 250 ms. After this time period, the voltage goes below the level set by R8 and R9, and the comparator U6 output goes low. This signals a valid receive carrier, which releases the muting signal (Q3 turns off, removing the short to ground from the output of U12).

Modem circuit. The modulation and demodulation of data are done by the CMOS modem IC U9, shown in Fig. 6-26. The modem IC generates its own clock using a built-in crystal (Y2) oscillator. The 1-MHz clock is also used to drive processor U3. Modem IC U9 has two sections: receive and transmit. Data bits received from the limiter are decoded and sent through RS232C drive U7 to the terminal. Data bits from the terminal are turned into digital-synthesized sine waves and sent to the transmit filter. The transmit enable (T-ENABLE) input is used to mute the carrier out of the modem for 2 seconds after the line has been seized.

The MOD-MODE control line from processor U3 is used to activate U9 for the local test mode. Note that S1, in the remote test (RT) mode, directly connects the data in from the terminal to the data going out to the terminal. This produces a condition known as *digital loopback* and is used as one of many standard test procedures for modem systems.

We will not go into modem system tests here. (Such tests are described in boring detail in the author's *Handbook of Data Communications* referenced at the beginning of this chapter.) However, so that you will understand the need for local/remote test circuits shown here, the following are brief examples of modem system tests.

Figures 6-27 and 6-28 show the basic test configurations for *local analog loopback* and *remote digital loopback*, respectively. In both cases, a signal (representing serial data marks) is sent to the modem for transmission and the output of the modem is monitored for reception of the signal.

In the local analog loopback test of Fig. 6-27, the modem transmitter and receiver are isolated from the telephone lines and the transmitter analog output is

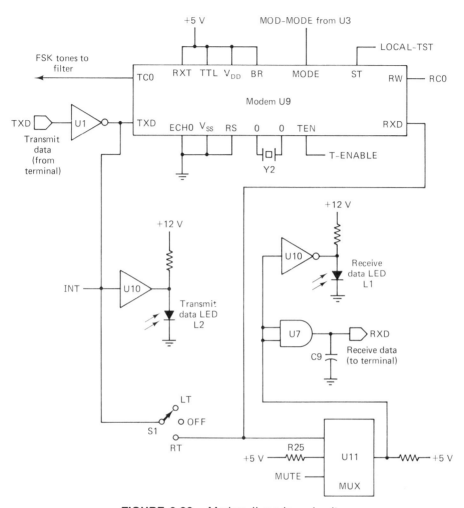

FIGURE 6-26. Modem II modem circuitry

internally looped to the receiver input. Typically, the local analog loopback test is performed at each modem in the system to check the local transmit/receive functions on the modem. Failure of the local analog loopback test indicates a faulty modem.

The remote digital loopback test of Fig. 6-28 checks out the ability of two (or more) modems to transmit and receive data. A successful completion of this test indicates that the telephone line between the modems is satisfactory (in addition to proving that the modems are good).

Getting back to the modem shown in Fig. 6-26, note that there is a special circuit involving a 2:1 CMOS multiplexer (part of U11) on the output of U9.

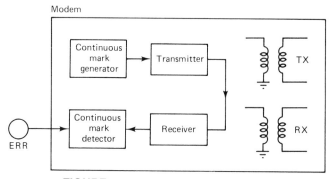

FIGURE 6-27. Local analog loopback test

Since the processor U3 echoes back valid programming data sent to it in the programming mode, these data bits must also pass to the terminal. However, U9 has a low output in the idle state. This effectively grounds the data from processor U3, so multiplexer U11 is used to insert a pull-up resistor R25 in place of the modem output. Processor U3 can easily pull this down and thus echo to the terminal.

Processor and tone dialing. The modem is controlled by U3, which reads the front-panel switches and monitors data from the computer. In this way, the modem may be configured to any of six operating modes (which are discussed thoroughly in the operating manual for the modem).

Upon power-up, U3 does an internal reset controlled by C2, as shown in Fig. 6-23. After initialization, the processor polls the status of bit O of port C (pin 8). If bit O of port C is low, U3 then examines the front-panel switches to see what mode to execute.

The programmer has three programmable I/O ports. Any port may be either an input or an output port, depending on the status of the internal program.

U4 is a parallel-loading tone-dial IC which has an enable pin connected to A0 of processor U3. When U4 is disabled, U3 uses A2 and A3 as switch inputs.

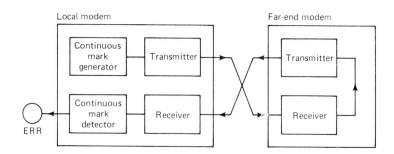

FIGURE 6-28. Remote digital loopback test

When tone-dialing, the switch inputs are programmed as outputs. U4 produces a tone pair which is determined by the binary inputs at A2 and A3. Crystal Y1 provides internal clocking. Capacitor C4 acts as a smoothing filter to remove harmonics. The dialing tones are coupled by C19 and amplified by U16.

6-3.2 Adjustments

The only adjustment required for the modem involves setting the level of the transmit output from modem U9 to the telephone line. Figure 6-29 shows the adjustment diagram.

With a 600-Ω resistor connected across the hybrid transformer T1, monitor the ring-to-tip output. Adjust R7 for an output of about 0.2 V(rms) (−10 dB) with a steady digital input at the TXD terminal, or pin 13 of U1, as shown in Fig. 6-29.

6-3.3 Troubleshooting

As in the case of the Modem I described in Sec. 6-2, troubleshooting for Modem II involves checking for signals at various points in the circuits. However, since Modem II has several operating controls and indicators, an evaluation of trouble symptoms can help pinpoint problems.

Again, Fig. 6-14 summarizes the check procedures for any modem. The following paragraphs go on to describe the checks of individual circuits, based on symptom evaluation and signal analysis.

Preliminary checks. The operating controls and indicators should be checked before you launch into troubleshooting on a modem (as is the case with any electronic device!). The following paragraphs describe some typical proce-

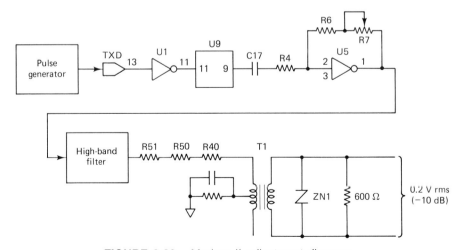

FIGURE 6-29. Modem II adjustment diagram

dures for Modem II. Figure 6-30 shows the circuits involved.

Make sure that POWER switch S4 is set to on and that power LED L6 is turned on.

Check the status of MODE switches S2 and S3. If S3 is in AUTO, calls can be placed and received automatically through the modem. If S3 is in MAN, S2 must be set to ANS or ORG to answer or originate calls.

Note that when S3 is in AUTO, the modem accepts a present number of rings before the modem automatically answers the telephone line call. The number of rings depends on the position of the jumper plug connected to U2.

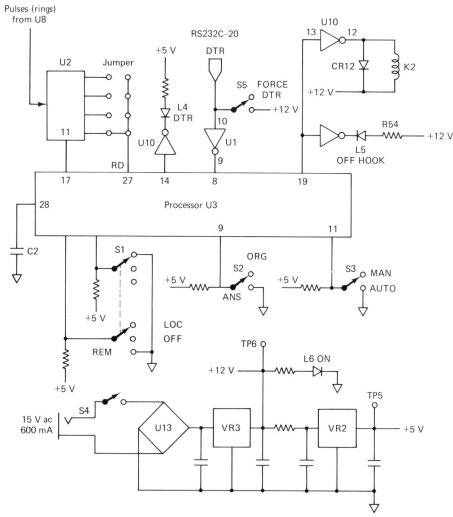

FIGURE 6-30. Preliminary check troubleshooting diagram

Check that TEST switch S1 is set to OFF. If S1 is set to LOC or REM, the modem goes into a remote or local test.

Check that the DTR indicator L4 is on, indicating that the data terminal is ready. If not, press FORCE DTR switch S5 and see if L4 turns on. If L4 turns on with S5, the problem is likely to be in the computer terminal rather than in the modem. Look for a high at pin 20 of the RS232C input when the terminals are assumed to be ready. If L4 does not turn on when S5 is pressed, suspect L4, U1, U3, or U10.

Check that OFF HOOK indicator L5 is on, after a preset number of rings, indicating that the modem has seized the telephone line. If not, check for a high at U3-19 and that relay K2 is actuated. If U3-19 is not high, suspect U3. If U3-19 is high, suspect U10. If L5 is on, but the line is not seized, suspect U10, CR12, or K2.

Keep in mind that U3-19 does not go high until U3 is reset and has received a ring detect (RD) signal from U2 and the ring-preset jumpers. U3 is reset after power-up (the time for reset is determined by C2 at U3-28). When U3 is reset, the reset line connected to U2-11 goes low, resetting U2 and placing U2 in a condition to count ring pulses from optocoupler U8. When the preset number of pulse (rings) are applied to U2 by U8, the RD signal is applied to U3 at pin 27, causing U3-19 to go high.

Ring circuits. Figure 6-31 shows the circuits involved.

If there appear to be problems in the ring circuits, try injecting a ring signal at the tip and ring inputs. Check that pin 5 of U8 pulses when the ring signal is applied. U8-5 is normally high because of the B+ through R2, but drops to low during each ring signal.

If there are no pulses at U8-5, check for a ring signal across CR4–CR7. If absent, suspect C53/R31. Then check for a d-c voltage across pins 1 and 2 of U8 during each ring signal. If absent, suspect Q2, CR3, and the associated circuit.

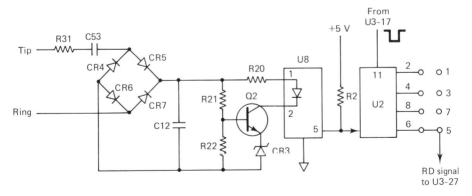

FIGURE 6-31. Ring-circuit troubleshooting diagram

If there is a d-c voltage at pins 1 and 2 of U8 during ring but no pulses at U8-5, suspect U8.

If there are pulses at U8-5, check that U2 counts each ring signal. The ring-detect (RD) signals should appear at pin 2 of U2 on the first ring, at pin 4 on the second ring, at pin 8 on the third ring, and at pin 6 on the fourth ring.

Keep in mind that U2-11 must be low (from U3-17) for U2 to count. If the count is not correct, suspect U2.

Also make sure that the RD signal passes to U3-27 through the correct jumper plugs or termnals. For example, if the jumper is connected across terminals 5 and 6 and there are only three rings, the RD signal does not reach U3-27.

Pulse-dial circuits. Figure 6-32 shows the circuits involved. The telephone line is pulse-dialed by signals from U3-18 (in response to commands applied to U3 from the computer terminal). The signals open and close relay K1, alternately shorting across R44/C22.

If the modem appears not to be pusle dialing properly when commanded by the terminal, check for pulses at U3-18. If absent, suspect U3 or the terminal.

If there are pulses at U3-18, check that relay K1 opens and closes for each pulse. If not, suspect U10, CR9, and K1. Also check that C22 is not shorted.

Also note that mute transistor Q4 shorts across the low filter input when pulse dialing (or programming U3). This prevents the large-voltage switching spikes (produced by opening and closing the telephone line) from damaging the filter circuits. The short across Q4 also prevents signals on the telephone line from passing through the filters during programming of U3 by the computer terminal.

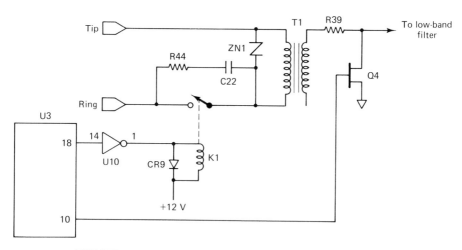

FIGURE 6-32. Pulse-dial-circuit troubleshooting diagram

Tone-dial circuits. Figure 6-33 shows the circuits involved. The tone pairs are produced by U4, under control of clock crystal Y1, in response to commands from U3 at port A (A0–A4, pins 20–24 of U3). The tones are applied to the hybrid transformer T1 at the junction of R40 and R50 through C19, R45, and amplifier U16.

If the modem appears not to be tone dialing properly when commanded by the terminal, check for tone pairs at U4-17. If absent, suspect U4, Y1, C4, R3, or the terminal.

If there are tone pairs at U4-17, trace the tone pairs through C19, R45, U16, and R40 to T1.

Transmit output. Figure 6-34 shows the circuits involved. If the modem appears not to be transmitting data to the telephone line properly when the terminal is being operated (but with good reception), check that TRANSMIT DATA indicator L2 is turned on. Also check for digital pulses at U9-11 and for transmit signals at U9-9.

If the pulses are absent at U9-11, the problem is probably in the terminal, although U1 and U3 could be at fault. Check for a T-ENABLE signal from U3-16 to U9-12.

Check for pulses at U1-13, U1-11, and U3-2. The pulses at U1-13 should

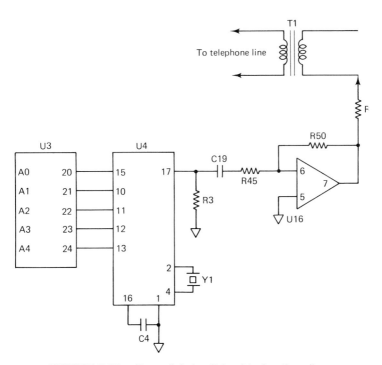

FIGURE 6-33. Tone-dial-circuit troubleshooting diagram

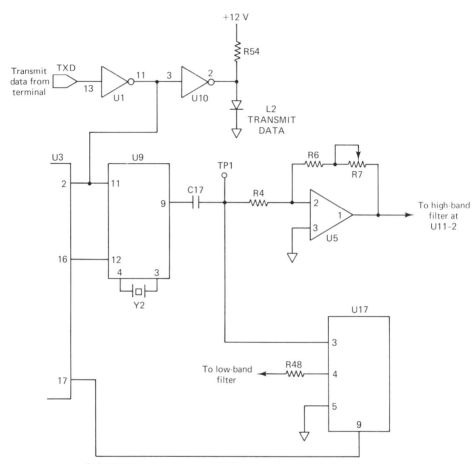

FIGURE 6-34. Transmit-output troubleshooting diagram

be the same as those from the terminal at the TXD connector. The pulses at U1-11 and U3-2 should be inverted from those at U1-13.

If the pulses are present at U1-11, but the TRANSMIT DATA indicator L2 does not turn on, suspect L2, R54, or U10 (pins 2 and 3).

If there are pulses at U9-11 but no transmission signals at U9-9, suspect U9, or possibly Y2. However, if the modem is passing received information from the telephone line to the terminal, Y2 is probably good.

Trace pulses from U9-9 through C17, (TP1), R4, U5, R6, and R7 to the high-band filter at U11-2. Note that when U17 is in the test mode (determined by the LOCAL TEST signal from U3-17 applied to U17-9) the transmitted signals are also applied to the input of the low-band filter network (through R48, which is usually connected to ground through U17-5).

High-band filter. Figure 6-35 shows the circuits involved. When troubleshooting the high-band filter circuits, trace the signals from input to output in the normal manner. During transmit, the high-band filter removes any pulse modulation from the transmission signal input (at U11-2) and produces pure sine waves at the output (U17-2 or TP4). During receive, the high-band filter removes any harmonics on noise present on the telephone line.

Note that the high-band filter is used for both transmit and receive, depending on the status of the MODE-FIL line (answer or originate) from U3-25 to U11-10 and U17-8.

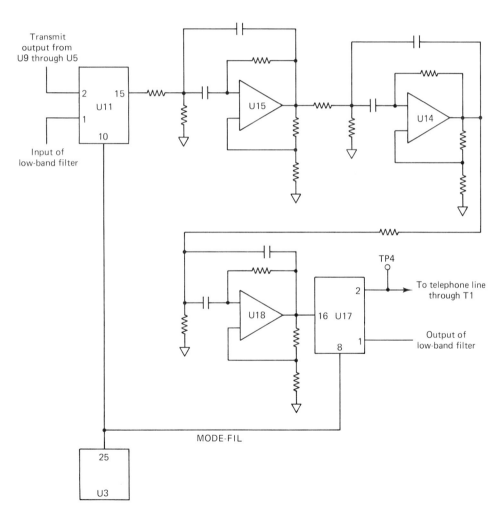

FIGURE 6-35. High-band-filter troubleshooting diagram

Low-band filter. Figure 6-36 shows the circuits involved. When troubleshooting the low-band filter circuits, trace the signals from input to output in the normal manner. The low-band filters operate as described for the high-band filters. Keep in mind that the low-band filter is also used for both transmit and receive, depending on the status of the MODE-FIL line (answer or originate) from U3-25 to U11-11 and U17-11.

Receive input. Figure 6-37 shows the circuits involved. If the modem appears not to be receiving data known to be present on the telephone line (but with good transmission), the problem can be in either the receive input circuits or the carrier-detect circuits (described next).

When troubleshooting the receive input circuits, trace the signals from the input at T1 through to the output at U16-1. Consider the following when troubleshooting the receive input circuits.

Make sure that Q4 is not turned on by a signal on the mute line connected to U3-10, except during pulse dialing or when U3 is being programmed by the terminal. If Q4 is turned on, the receive input is shorted to ground.

Most of the gain for the high-band and low-band filter circuits is provided by U16 during receive. If there are received signals on the line but they do not produce a valid carrier (CARRIER DETECT indicator L3 is not on), suspect U16. Also look for a partially shorted CR8.

Carrier-detect circuits. Figure 6-38 shows the circuits involved. If the problem is one of good transmission but no reception (with the receive input circuits apparently good), try injecting a signal at the primary of T1 (Fig. 6-37) or receive input, and check that the CARRIER DETECT indicator L3 turns on. Also check that there is a signal at the output to the terminal (at the RXD connector) and that the RECEIVE DATA indicator L1 turns on.

If L3 turns on but there is no signal at the RXD connector (or U7-11), and L1 does not turn on, suspect U12, U9, U7, U10, or U11.

It is also possible that U11 has been switched by a mute signal from U3-10. This should occur only during programming (or during a pulse dial). If U11 is switched, the receive output from the modem at U9-7 is replaced by R25, and no receive pulses are sent to the terminal.

Also note that when S1 is in REM (remote test), data bits to be transmitted from the terminal (and applied to the modem at U9-11) are also applied to the RXD connector through U11-5 and U7.

If L3 does not turn on, make sure that there is a proper signal at C21 (filter output). Typically, this signal is a sine wave of about 5 or 6 V. If there is no signal at C21, suspect the filter circuits.

If there is a proper signal at C21 but L3 does not turn on, suspect L3, U5, U6, CR1, CR2, and the related circuit components.

Keep in mind that the output at U6-1 does not go low (to turn Q3 off and permit passage of received signals from U12 to U9) for about 250 ms after L3

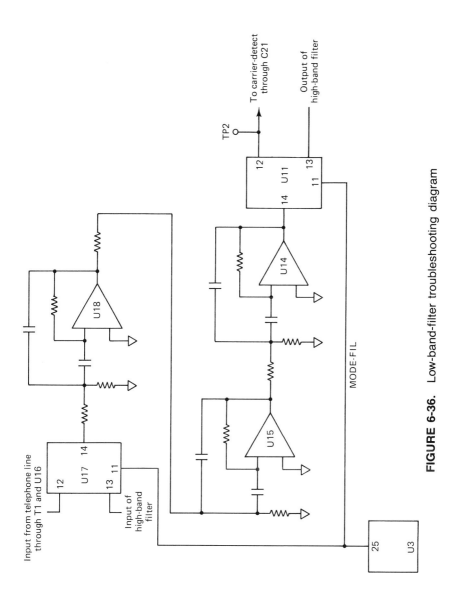

FIGURE 6-36. Low-band-filter troubleshooting diagram

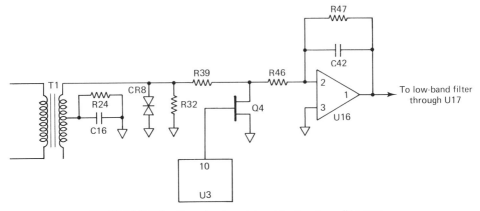

FIGURE 6-37. Receive-input troubleshooting diagram

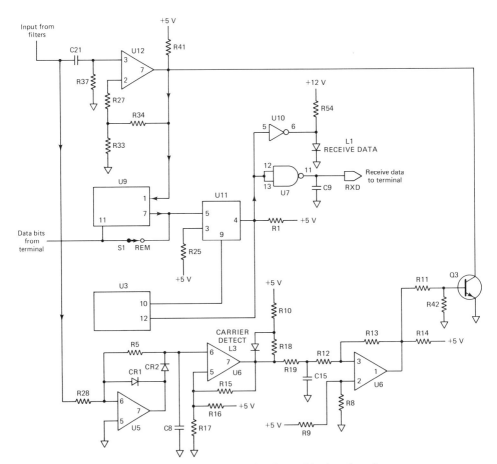

FIGURE 6-38. Carrier-detect-circuit troubleshooting diagram

turns on. This is done by discharging C15 through R19 when pin 7 of U6 goes low (and L3 turns on). During the 250 ms interval, with U6-1 high, Q3 turns on and prevents signals at U12-7 from reaching U9-1. If C15 does not discharge properly, U6-1 may remain high and cut off signals to U9-1.

If L3 is turned on, check the collector of Q3. The collector should be at some voltage above ground (as should U12-7). If not, suspect U6, C15, and the related circuits.

6-4. TELEPHONE INTERFACE II (ACOUSTIC COUPLER) TROUBLESHOOTING

This section describes troubleshooting for a typical telephone interface or acoustic coupler (whichever term you prefer). The Radio Shack Telephone Interface II is selected as the example. Again, we start with a description of the interface circuits.

6-4.1 Circuit Descriptions

The Telephone Interface II is an asynchronous modem designed to operate up to 300 baud over a dial-up telephone network, using FSK modulation. Figure 6-39 shows the major components and controls. The interface is switch-selectable to operate in either Originate or Answer modes, using RS232C signals. The full specifications for the interface are described in the service literature. The interface operates at the following voltage levels (which conform to RS232C):

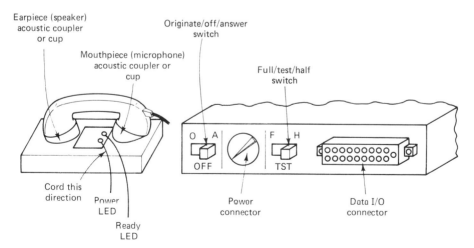

FIGURE 6-39. Major components and operation controls of Radio Shack Telephone Interface II

Inputs:
    Mark (off) −3 to −25 V
    Space (on) +3 to +25 V
Outputs:
    Mark (off) −8 V
    Space (on) +10 V

Basic operating principles. Figure 6-40 shows the functional relationship of the interface to the telephone system and computer. Figure 6-41 is a block diagram of the interface. The following paragraphs describe the basic operating principles of the interface during various modes of operation.

When power is first applied. If power is applied to the telephone interface in the originate mode before the telephone handset is inserted in the microphone and speaker cups, the output data state of the receiver is forced to the mark state by the absence of a valid carrier. The transmitter is also disabled (held off) until a valid carrier is received.

Normal originate and answer modes. Figure 6-40 shows the normal originate/answer mode of operation, where the computer on the left (originate) initiates a call to communicate to the computer on the right (answer) via modems and the dial-up telephone line.

The first step in this mode is to communicate with the remote computer (or possibly a time-sharing service) using the dial telephone in the normal manner. Once this is complete, a signal is heard in the local or originate telephone receiver. This signal is the *answer mark frequency* (2225 Hz) from the remote or

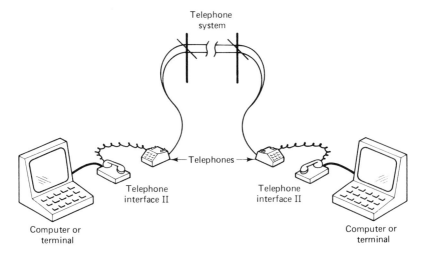

FIGURE 6-40. Functional relationship of Telephone Interface II to the telephone system and computer

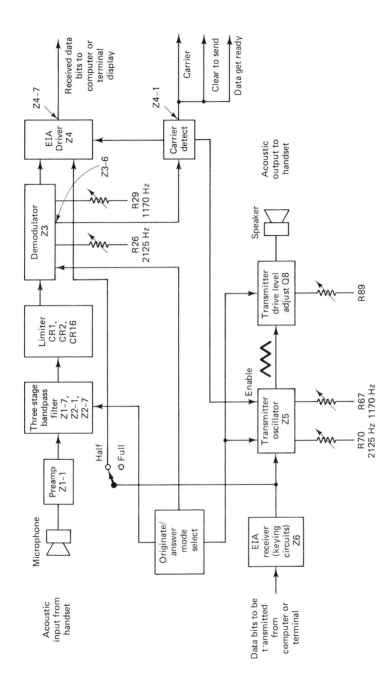

FIGURE 6-41. Radio Shack Telephone Interface II block diagram

answer modem. When the telephone handset is properly inserted into the microphone and speaker cups of the originate modem, the received signal is coupled to the microphone. Simultaneously, ambient noise is greatly attenuated by the close-fitting microphone cup assembly.

The signal from the microphone is amplified and applied to a limiting circuit (Fig. 6-41) that produces sharp leading and trailing edges corresponding to the zero crossings of the received signal. The demodulator determines the frequency of the incoming signal as either *mark* (2225 Hz) or space (2025 Hz). Approximately 3 seconds after a valid signal from the answer modem, the carrier detector in the originate modem enables the originate transmitter.

The transmitter output drives the speaker in the telephone handset microphone cap. The originate modem then sends an *originate mark frequency* (1270 Hz) to the dialed answer modem and computer. Data transmission then proceeds between the modems by frequency-shift keying the transmitters in accordance with the digital input signal to the corresponding transmitters (from the computers, terminals, etc.).

If the modem is in the *half-duplex* operating mode, the data bits applied to the transmitter are also connected to the receiver output, so that the computer prints out both the data received from the remote computer, plus the local transmitted data applied to the local transmitter input. In half-duplex, the receiver output corresponds to either the transmitter input signal or the received signal, simultaneously. This can result in garbling of the data.

In the *full-duplex* operating mode, the transmitter operates independently. Data transmission and reception take place simultaneously. Only the received data bits are displayed.

When data transmission is complete, the telephone handset is removed from the modem and placed on-hook. This terminates communication. The carrier detector circuit senses interruption of a valid carrier. This returns the receiver output to *mark hold*, turns the READY indicator off, and disables the transmitter. The modem is then returned to the idle state.

Receiver operation. The receiver functions to amplify the low-level signal from the microphone to the signal level required by the demodulator. The amplifier includes a three-stage bandpass filter to attenuate noise and extraneous signals outside the desired passband.

The bandwidth of the filter is approximately 400 Hz, centerd at 2125 Hz in originate and 1170 Hz in answer. The receiver consists of the preamp stage, the bandpass filter, a limiter circuit, a frequency demodulator, and an output driver.

Preamp stage. Figure 6-42 shows the circuits involved. Two configurations of the preamp are available. One configuration uses a capacitive-type microphone. The other configuration uses a ceramic microphone. When the capacitive microphone is used, the preamp gain is about 80 dB. The preamp gain is approximately 22 dB with the ceramic microphone. As shown by the waveform in Fig. 6-42, the output from Z1-1 is approximatley 70 mV peak to peak.

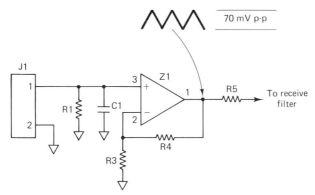

FIGURE 6-42. Preamp stage

Receive filter. Figure 6-43 shows the circuits involved. The three-stage bandpass filter consists of Z1-7, Z2-1, and Z2-7 and their associated circuit components. The flat bandpass characteristic is achieved by staggering the center frequencies of each stage, as shown in Fig. 6-43. The composite gain of the three stages is 44 dB at the center frequency.

Note that when S1 is set to 0 (originate), Q1, Q2, and Q3 are all turned on. This alters the characteristics of all three stages in the receiver filter simultaneously. As shown by the waveform in Fig. 6-43, the output from Z2-7 is approximately 10 V peak to peak.

Limiter and demodulator. Figure 6-44 shows the circuits involved.

The limiter circuit consists essentially of diodes CR1 and CR2, which create a deadband for signal threshold. The signal must be greater than −50 dBm. Diode CR16 limits the signal applied at the demodulator to about 2.5 V, as shown by the waveform.

The demodulator Z3 is an IC especially designed for FSK modem applications. The internal circuit consists of a basic PLL for tracking an input signal within the passband, a quadrature phase-detector which provides carrier detection, and an FSK voltage comparator which provides FSK demodulation.

External components are used to set center frequency, bandwidth, and output delay. The free-running frequency is set by C13 and the current drawn from pin 11. In the answer mode, resistors R29, R30, and R32 set the free-running frequency to 1170 Hz by adjustment of R29. In originate mode, Z6 is in saturation, thus paralleling R26, R27, and R28 with the answer resistors, and R26 adjusts the frequency to 2125 Hz.

Because the PLL error voltage is proportional to incoming frequency, R26 and R29 are, in effect, *received data symmetry* adjustments. R33 and C16 form the loop filter. R34 and C17 smooth data, and R39/C15 smooth carrier-detect signal.

The carrier-detect output of demodulator Z3-6 is high when the PLL is locked and low when a valid carrier is absent. Carrier-detect on-and off-time

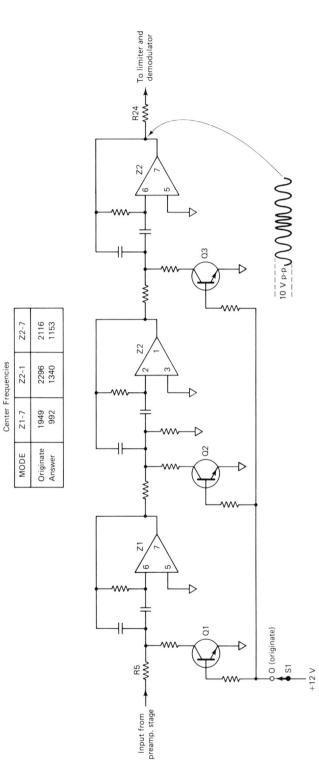

FIGURE 6-43. Receive filter

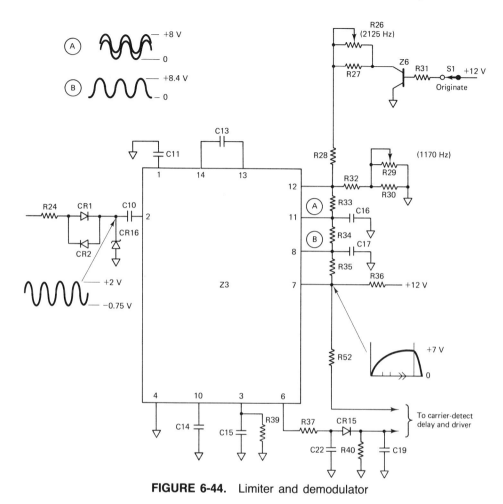

FIGURE 6-44. Limiter and demodulator

delays of approximately 0.7 and 1.2s are provided by circuits (R37, R40, C19, C22, and CR15) external to Z3.

Carrier-detect delay and driver. Figure 6-45 shows the circuits involved. IC Z4 contains circuits that provide RS232C level outputs (Received Data) to the computer or terminal (at pin 7). Z4 also provides Clear to Send, Data Set Ready, and Carrier Detect signals to the computer or terminal (at pin 1). Note that READY indicator LD1 turns on when the signals are available. Also, note that the pin numbers on connector J4 correspond to the standard RS232C configuration shown in Fig. 6-2.

When S2 is set to H (half-duplex), data bits from the terminal (normally applied to the transmitter) are also applied to the driver Z4 (at pin 6, through CR10 and S2). These data bits are mixed with those from Z3 and are applied to the terminal as Received Data (at J4-3).

228   Modem and Telephone Interface Troubleshooting

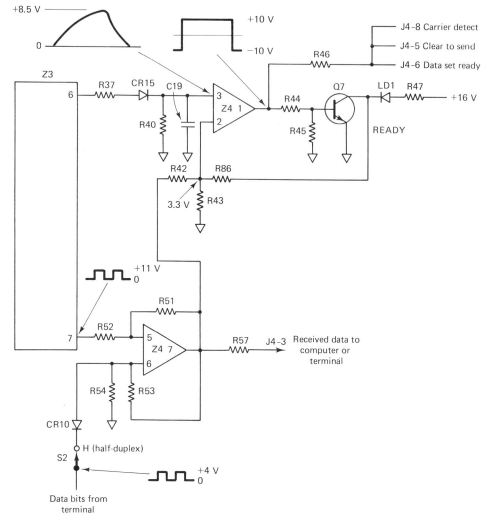

FIGURE 6-45. Carrier-detect delay and driver

Transmitter operation. The transmitter converts digital input signals to tones for transmission over the telephone line to the remote modem. In the originate mode, a mark is represented by a tone with a frequency of 1270 Hz, and a space by a frequency of 1070 Hz. In the answer mode, the mark and space tones are 2225 and 2025 Hz, respectively.

In the absence of a valid carrier, the transmitter is disabled when the modem is in the originate mode. In the answer mode, the transmitter sends a continuous mark signal, a few seconds after power is applied (power up). The transmitter consists essentially of an oscillator, keying, and enabling circuits.

Transmitter circuits. Figure 6-46 shows the circuits involved. The oscillator Z5 is a variable-frequency IC which provides a signal of triangular shape to the speaker. The frequency is determined by C21 and the resistance connected to Z5-4 (the lower the resistance, the higher the frequency).

In the originate mode, the base of Z6-8 is negative and resistors R69, R70, and R79 are effectively disconnected from Z5-4, leaving R66, R67, and R68 to determine frequency. Frequency is set by R67 to 1170 Hz with a 50% duty cycle square wave applied to J4-2. This simulates an alternate mark-space keying pattern.

In the answer mode, Z6-7 is in saturation, causing the branch containing R69, R70, and R79 to be connected at Z5-4. The answer frequency is then adjusted to 2125 Hz by R70.

Drive level to the speaker is set by R78 in answer (since Q8 is turned on). In originate, the speaker drive is set by R78, R75, and R89 (since Q8 is turned off).

The keying circuits consist of two sections of transistor array Z6 and associated components. With no signal applied to J4-2 from the terminal, or a negative voltage applied, the Z6 input transistor is biased off by R63 and CR12, which clamp Z6-13 one diode drop below ground. Z6-14 goes high, causing Z6-1 to be in saturation and effectively shorting out R66. This causes the oscillator frequency to be high (or in mark) in either the originate or answer mode. A positive input at J4-2 reverses these conditions, causing R66 to be added to Z5-4. This results in a lower oscillator frequency (or a space).

Approximately +5 V bias is required at Z5-5 for proper operation of the oscillator. Pin 5 of Z5 is therefore used to enable or disable the transmitter circuits. In the answer mode, Z6-12 is in saturation due to conduction through CR13, while Z6-9 is effectively an open circuit. This causes the transmitter to be continuously enabled in answer.

In originate, CR13 is reverse biased through R87, Z6-12 turns off, and Z6-9 turns on, shorting Z5-5 to ground. This disables the transmitter until a valid-carrier signal from the carrier-detect circuits (Fig. 6-45) is applied to CR14. With CR14 forward biased, Z6-12 turns on and Z6-9 turns off, restoring Z5-5 to about +5 V. This turns on the transmitter as long as a valid-carrier signal is present.

Power supply. Figure 6-47 shows the circuits involved. Power is supplied at 24 V ac from a wall-mounted UL-approved transformer. A bridge rectifier CR4–CR7 generates approximately +30 V across C18. The 30 V is divided by R49, R50, and CR8/CR9 to provide +16, +12, and −10 V. The common point between CR8 and CR9 establishes power and signal ground.

G-4.2 Test/Adjustment/Troubleshooting

The following paragraphs describe combined test and troubleshooting procedures for the interface. The test procedures covered here establish that all of the circuits

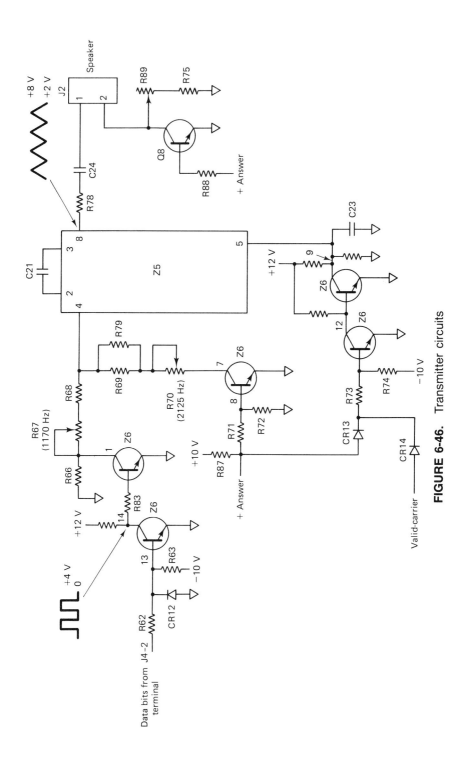

FIGURE 6-46. Transmitter circuits

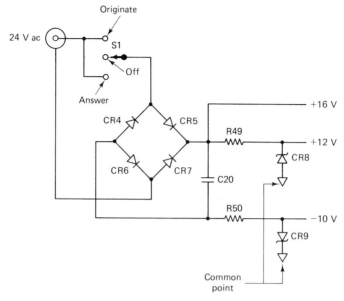

FIGURE 6-47. Power supply

are functioning normally (if the interface performs properly) or pinpoints malfunctioning circuits (if the interface fails to pass one or more of the tests).

Basic test connections. Figure 6-48 shows a simple way to check out the interface using the internal test feature. Switch S2 in the interface is set to TEST. This causes the transmitter to produce the band of frequencies normally received by the modem. The transmitter output is then fed back to the input through a 1000:1 attenuator. The output is obtained from the switch side of J2 and, after attenuation, is connected to the input side of J1. Note that J1 is shown in Fig. 6-42 and is the microphone input (produced by acoustic output from the telephone earpiece). J2, shown in Fig. 6-46, is the speaker output (and produces acoustic output to the telephone mouthpiece or microphone). Switch S3 in Fig. 6-48 facilitates check of the carrier-detect circuit timing.

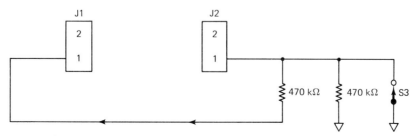

FIGURE 6-48. Basic test connections for telephone interface

Test conditions. The test conditions used to obtain the waveforms and voltage levels shown on the schematic diagrams (Figs. 6-42 through 6-47) are as follows:

1. The output connected to the input as shown in Fig. 6-48.
2. The transmitter modulated by 300-baud alternating signal (square wave of 150 Hz, 10 V peak to peak) applied to J4-2 (Fig. 6-46).
3. Line voltage of 115 V.
4. Switch S1 in interface set to 0 (originate).

Note that the output at Z5-8 (Fig. 6-46), and thus the input at Z1-1 (Fig. 6-42), is triangular under these test conditions. The triangular waveform is very low in harmonic content and becomes a sine wave when passed through the speaker.

Transmitter frequencies. Using the test connections of Fig. 6-48, connect a frequency counter and oscillocope to J2-1. Verify the waveform at Z5-8 and switch the interface to answer (S1 to A). The READY indicator LD1 (Fig. 6-45) should remain on and the frequency counter should read 1170 Hz (the average of 1070 and 1270 Hz). If necessary, adjust R67.

Disconnect the generator from J4-2 and check that the frequency rises to within 5 Hz of 1270 Hz (mark).

Connect J4-2 high by connecting to +12 V, and check that the frequency lowers to within 5 Hz of 1070 Hz (space).

Switch back to originate and reconnect the generator to J4-2. The frequency counter should read 2125 Hz. If necessary, adjust R70.

Disconnect the generator from J4-2 and check that the frequency rises to 2225 Hz (mark). [Ground Z6-12 (Fig. 6-46) to enable Z5 in the absence of a carrier in the originate mode.]

Connect J4-2 high by connecting to +12 V and check that the frequency lowers to 2025 Hz (space). Note that it is important that any adjustments be made in the sequence indicated.

Received data symmetry. With the transmitter keyed by a 50% duty cycle square wave at J4-2, the received data should also be a 50% square wave. With the basic test connections as shown in Fig. 6-48 and READY indicator LD1 on, connect an oscilloscope to J4-3 (Fig. 6-45), switch to answer, and adjust the oscilloscope variable time base so that one cycle (2 bits) of received data fills the screen (typically 10 divisions on the screen). The mark-to-space transition should occur at exactly 5 divisions. If not, adjust R29 (Fig. 6-44). Switch to originate and adjust R26 as required. Again, the sequence of adjustments is important.

Carrier-detect timing. Figure 6-49 shows timing diagrams for three points in the carrier-detect circuit (Fig. 6-45). These waveforms can be verified

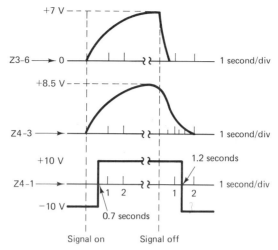

FIGURE 6-49. Timing diagram for carrier-detect circuits

by setting the oscilloscope time base to 1 second/div (or faster if desired) and operating switch S3 at the attenuator (Fig. 6-48).

As an example, connect the oscilloscope to Z4-1 and close S3 to short out the attenuated signal. Z4-1 should be approximately −8 V.

When the beam crosses a major division, open S3 to switch on the input signal. Z4-1 should go high (+10 V) in approximately 0.7 seconds.

Again close S2 to short out the signal, and note that Z4-1 goes low in approximately 1.2 seconds.

If the waveforms shown in Fig. 6-49 appear to be normal, the preamp, receive filter, demodulator, and carrier-detect circuits can be presumed good.

Receive filter gain and frequency response. As shown in Fig. 6-43, the receive filter is made up of three stages. The stages are stagger-tuned so that the overall 3-dB bandwidth is 400 Hz, centered about 1170 or 2125 Hz, depending on the answer/originate mode.

When testing the receive filter for gain and frequency response, note that there may be some slight variation in signal amplitude between a mark and a space, as observed at Z2-7 (even though the transmitter is on the correct frequency). However, a difference in mark/space greater than 20% can indicate trouble in the receive filter circuits (which often shows up as erratic operation of the modem).

Also, note that with a 70-mV peak to peak input to the receive filter, the final or overall output should be about 10 V peak to peak, even though there are three stages (each with a gain of about 9×). This is because of the stagger tuning.

The following steps describe how to check the first stage of the receive filter. The same procedure can be applied to the remaining stages.

Inject a sine wave at J1-1, and set the oscilloscope for *x-y* display, as shown in Fig. 6-50.

Connect the *x*-axis to the first filter stage input (Z1-1) and the *y*-axis to the filter stage output (Z1-7).

Vary the sine-wave signal across the center frequencies shown in the table of Fig. 6-43 (1949 Hz for originate; 992 Hz for answer).

The oscilloscope should show an elliptical pattern with the major axis in the second and fourth quadrants. At the exact center frequency, the ellipse should close to a straight line in the same quadrants.

The center frequencies should be within 20 Hz of the values shown in Fig. 6-43. If not, suspect the receive filter stage.

Note that the input signal amplitude must be kept to a level that does not saturate the stage under test.

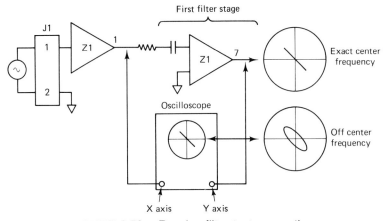

FIGURE 6-50. Receive filter test connections

INDEX

A

ACIA 185
Acoustic coupler:
 circuits 221
 troubleshooting 229
Audio:
 circuits 9
 tests 44, 49, 90

B

Base unit:
 adjustments 145
 circuits 122
Battery voltage 3, 20

C

Call button circuits 130
Call signal circuits 136
Call-waiting circuits 141
CCITT V.24 182

Coding, digital 12
Cord tests 27, 58, 74, 80, 95
Corded telephone tests 23
Cordless telephone 10
 circuits 120
 channel frequencies 38
 tests 24, 36, 67, 154

D

DCE 185
Dial circuits 8
 tests 85, 107
Dial decoder 22
Dial signal circuits 138
Dial systems 4
Dial tests 29, 60
Dial-tone circuits 136
 tests 96, 102
Digital coding 12
 tests 116
DTE 185
DTMF 5

235

DTMF (*Cont.*)
   tests 89

E

Electronic telephone 7

F

Flash circuits 141
Frequency tests, cordless telephone 45

G

Guardtones 12
   tests 42

K

Keystroke tone 136

L

Level tests 26
Line:
   circuits 126
   interface (modem) 189
   simulator 22
   tests 69, 79
Loop:
   resistance 3
   tests 74

M

Marks and spaces 188
Modem:
   adjustments 195, 211
   circuits 181, 189, 201
   troubleshooting 195, 211
Modulation tests 44, 49

O

Off-hook impedance 3

P

Pilot signal 12
   detector 14
   tests 42
Pilot-tone circuits 122
Portable unit:
   adjustments 149
   troubleshooting 133
Privacy, cordless telephone 11
Pulse dial 4, 214
   circuits 125
   cordless telephone 14
   tests 86, 108
Pushbutton dial 4

R

Ring and tip 4
Ring circuits:
   tests 82
   troubleshooting 99
Ring detection circuits 128
Ringer circuit 7
Ringer equivalance numbers (REN) 33, 73
Ringer tests 81
Ring frequencies 12
Ring signal:
   generator 13
   tests 48
Ring tests 31, 63, 71, 96
Ring voltage 4, 20
   generator 22
Rotary dial 4
   tests 85
RS232C 182
RS422, RS423 184

S

Safety, telephone service 18
Security code circuits 122
Spaces and marks 188

T

Telephone interface:
　circuits 221
　tests 229
Teletype interface 188
Tip and ring 4
Tone dial 4, 15, 215
　circuits 136
　tests 89, 109
Touch tone 4
　tests 89
Troubleshooting, basic 75
Two-wire circuit 2

V

Voice-circuit troubleshooting 111
Voice level 65
　tests 33
Voice quality 66
　tests 35
Voice tests 97